Stephan J. Nix

# Urban Stormwater Modeling and Simulation

LEWIS PUBLISHERS
Boca Raton Ann Arbor London Tokyo

**Library of Congress Cataloging-in-Publication Data**

Nix, Stephan J.
Urban stormwater modeling and simulation / Stephan J. Nix.
p. cm.
Includes bibliographical references and index.
ISBN 0-87371-527-6
1. Urban runoff—Mathematical models. 2. Urban runoff—Computer simulation. I. Title.
TD657.N57 1994
628′.21′015118—dc20 93-45680
CIP



Direct all inquiries to CRC Press, Inc., 2000 Corporate Blvd. N.W., Boca Raton, Florida 33431.


Lewis Publishers is an imprint of CRC Press


International Standard Book Number 0-87371-527-6
Library of Congress Card Number 93-45680
Printed in the United States of America 1 2 3 4 5 6 7 8 9 0
Printed on acid-free paper

# AUTHOR

**Stephan J. Nix, Ph.D.,** has been Associate Professor and Director of Environmental Engineering at Syracuse University since 1983. Dr. Nix received a B.S. in engineering science, an M.E. in environmental engineering, and a Ph.D. in environmental engineering from the University of Florida, Gainesville.

Dr. Nix has nearly 20 years of experience in urban stormwater management, hydrologic and water quality modeling, and computer applications. He has authored or co-authored over 25 papers and 2 books, including *Control and Treatment of Combined Sewer Overflows*, published by Van Nostrand-Reinhold, and *Urban Stormwater Modeling And Simulation*, soon to be published by Lewis Publishers. He has been consultant and advisor to several engineering firms and governmental agencies in the areas of urban stormwater modeling, economic analysis, and water quality.

Dr. Nix is a member of the American Society of Civil Engineers, Water Environment Federation, and the American Water Resources Association (AWRA). He has held several offices, most notably in AWRA, as president of the New York Section, Middle Atlantic District Director, and Secretary for the national organization.

# PREFACE

I have always had a love–hate relationship with computer models in general and urban stormwater models in particular. I love the freedom to experiment they allow and always enjoy getting one to work well. However, I have always detested the slightest hint of mystique surrounding them — a mystique that turns some people away out of fear and has some putting too much faith in them. I have built a few models, looked at the insides of many others, and taught many people how to use them. Some models are pretty clever and many are very large, but for the most part there is not much that should cause fear or awe. However, I do not want to say this too strongly or the reader might be inclined to disregard this book.

A book on urban stormwater modeling is needed for several reasons. First, there is renewed interest in urban stormwater and its management, especially in terms of water quality. As conventional point sources come under control, it is becoming more and more evident that diffuse sources of pollution must be controlled. Second, the dynamic nature of urban runoff requires considerable analysis. This often involves a computer model. Third, there are few, if any, works on urban stormwater modeling that go into enough detail to be useful while at the same time remaining focused on the *practical* issues.

The goal of this book is to provide a technical base for urban stormwater model users and a modeling approach within which to operate. This is done by describing and explaining the inner workings of a few of the more popular urban stormwater models and discussing how to use them in practical terms. The material is not advanced, but I think it can be useful. I hope that newcomers will, with this book's help, develop the ability to work effectively and efficiently with urban stormwater models and that modeling veterans will develop a little more savvy.

The text is organized into five chapters. Chapter 1 is a brief overview of urban runoff problems and how computer modeling plays a role in solving those problems. Chapter 2 reviews and explains the basic theoretical and numerical formulations of a representative set of stormwater models. Chapter 3 demonstrates how to construct a credible model of an urban watershed by thorough analysis of the modeling objectives and the processes of verification and validation. Chapter 4 focuses on using models to make predictions about stormwater runoff and pollution discharges from urban watersheds. Chapter 5 critiques some of the current stormwater modeling techniques by looking at seven common "sins" and provides some suggestions for future directions in the craft. Small, concise example applications are used throughout the book to enhance the discussion. An Appendix containing an annotated input data guide for the most popular urban runoff model, the EPA Storm Water Management Model, is also included.

This book, modest as it is, is the result of more than a decade of university teaching, consulting, and my own research in the field. I have learned from mentors, colleagues, and students. Some of their ideas and insights have certainly molded and contributed to what I call my own. I am deeply grateful to all of them.

# TABLE OF CONTENTS

# 1 THE URBAN RUNOFF PROBLEM AND MODELING

## 1.1 OVERVIEW

Precipitation falling over an urban watershed passes through an enormously complex hydrologic and hydraulic system. As it moves through this system it is concentrated into larger and larger flow streams and picks up a wide variety of pollutants in the process. An urbanized area is, by definition, an area of concentrated human activity. With this activity comes an increase in runoff volumes and flow rates due to the covering of much of the surface with impervious materials (concrete, asphalt, etc.). In addition, such a concentration of human activity can only be maintained by a large influx of a great variety of materials. Subsequently, there is a high concentration and diversity of waste materials. Some of this waste is transported from the urban area by stormwater runoff to receiving waters. This transport process is very efficient since urban areas have elaborate drainage systems to remove runoff quickly. Essentially, an urban watershed differs from its natural counterpart in that there are larger, more diverse waste discharges and a reduction in the watershed's ability to act as a buffer; physically, biologically, or chemically. Figure 1 presents a simplified picture of the urban watershed and the transport of water and pollutants.

Engineers have viewed urban stormwater with different perspectives over the years. In earlier times, the concern was for flood control and removing runoff as expeditiously as possible. In more recent times, the cross purposes of removing runoff from streets and parking lots and yet not overwhelming receiving waters led to the notion of comprehensive stormwater management. Today, engineers and planners are faced not only with the control and management of runoff quantity but the maintenance of water quality as well.

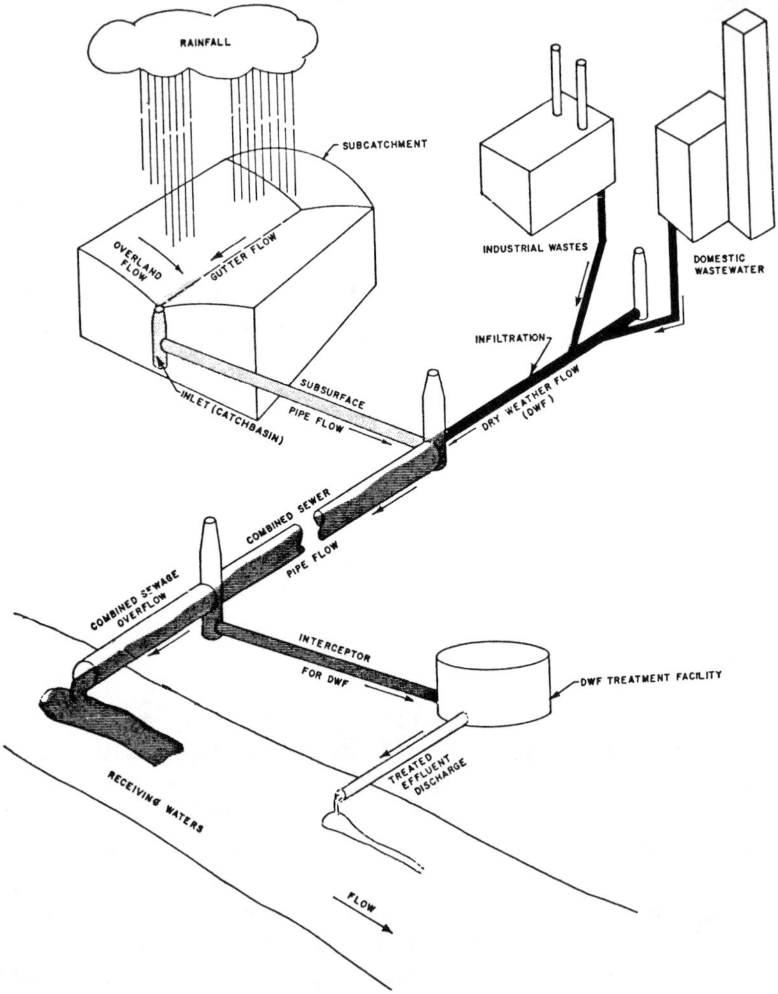


**Figure 1. Simplified schematic of the urban watershed (from Metcalf and Eddy, Inc. et al., 1971).**

Stormwater quantity problems remain a high priority in many localities. However, interest in urban stormwater quality has swelled as a result of the 1987 Water Quality Act. This act, the latest in a series of legislation that began with the 1972 Clean Water Act (the cause of an earlier period of intense interest), outlines a permitting system to regulate stormwater discharges from medium and large municipal storm sewer systems, current holders of NPDES (National Pollutant Discharge Elimination System) permits, and a host of industrial activities. The EPA published regulations in November 1990 to elucidate the permitting system (Federal Register, 1990). The result is that a good portion of the urban stormwater discharges in the U.S. are to be handled, from a regulatory perspective, as "point" sources of pollution (see Section 1.4).

Water resources engineers were among the first to use the computer to analyze large problems. Watershed models have been used since the late 1950s. However, the appearance of the Stanford Watershed Model in the mid-1960s marked the beginning of their widespread use (Crawford and Linsley, 1966). Models focusing on urban runoff and drainage have been around nearly as long. Urban models capable of simulating stormwater quality came along a little later as environmental concerns arose in the late 1960s and early 1970s. Appearing in 1971, the U.S. Environmental Protection Agency Storm Water Management Model with its ability to simulate flow and pollutants through complex urban watersheds was one of the first of these models and it remains preeminent today (Metcalf and Eddy, Inc., 1971).

This book will focus on the use of mathematical models to simulate the quantity and quality of stormwater runoff from urban watersheds. This chapter will briefly review those problems and discuss the role models play in developing and executing simulation experiments to help provide solutions. Subsequent chapters will explore several modeling software packages and how to use these packages to develop credible models and discuss the important aspects of conducting the simulation experiments.

There are many books from which to learn more about urban runoff and the problems associated with it. Three books are recommended here for further reading. One is a rather brief text by Whipple et al. (1983) that discusses the gamut of problems associated with urban runoff. A second is a more detailed book by Walesh (1989). Another by Moffa et al. (1990) focuses on the special urban runoff problem of combined sewer overflows. There are others, but the reader will find these to be concise, practical, and good starting points.

## 1.2 URBAN RUNOFF QUANTITY PROBLEMS

Rain falling over an urban watershed will strike either a pervious surface (i.e., a surface which will allow water to infiltrate to the subsurface) or an impervious surface (i.e., a surface which will not allow water to infiltrate). On pervious surfaces most of the rainfall infiltrates to the subsurface and some remains as surface runoff. A portion of the infiltrated water may take a slow subterranean path to a surface stream. On impervious surfaces nearly all rainfall becomes surface runoff. Surface runoff from both surfaces finds its way to channels and streams. Urban drainage systems speed this process along.

Urban watersheds are, of course, characterized by impervious surfaces and efficient drainage systems. The increased volumes and flow rates of runoff produced under these conditions have a number of deleterious effects, including the following:

- *Flooding*. Developed areas and their drainage systems are usually very good at discharging runoff — so much so that they cause flooding at downstream locations. The result is localized flooding. On the other hand, older or inadequately designed drainage systems can themselves be overwhelmed by runoff from increased urbanization. The result is flooding at the "headwater" of the drainage system.
- *Stream Erosion*. The increased runoff accompanying urban development increases the bed load, or sediment-carrying, capacity of a stream (Whipple et al., 1983). This diminishes the integrity of the stream bed and stream banks. In addition, the sediment load carried by the stream can accumulate at downstream points where the flow characteristics (e.g., velocity) are such that the bed load capacity is reduced.
- *Habitat Destruction*. A stream ecosystem is a delicate balance between all of its biological components. It is also in a precarious equilibrium with the physical environment. Increased runoff to a stream changes this balance and can threaten the established ecosystem. Increased thermal loads normally associated with urban runoff can also disrupt sensitive ecosystems.

## 1.3 URBAN RUNOFF QUALITY PROBLEMS

As rainfall moves through the atmosphere it washes out air pollutants and carries them to the surface. Upon reaching the ground it will dislodge some particles (mostly soil on pervious surfaces, a wide variety of dust and debris on impervious surfaces) and dissolve other materials. Surface runoff carries the particles dislodged by the initial precipitation impact, other particles dislodged by the movement of the runoff itself, and a variety of dissolved materials to drainage systems and watercourses. In some cases, the infiltrated water will threaten aquifers with a variety of pollutants.

The pollutants carried from the watershed surface come from a number of sources such as

- Transportation
- Industrial activities
- Decaying vegetation
- Soil erosion
- Animals
- Fertilizer/pesticide application
- Deicing agents
- Dryfall
- General litter

Pollutants may also be contributed by the watershed's drainage system. Such a system may consist of natural or artificial channels as well as sewerage. The sewer system may be designed exclusively for stormwater flows, in which case it is known as a *separate sewer system*, or designed to carry both stormwater and sanitary sewage flows, in which case it is known as a *combined sewer system*. In natural channels erosion can produce significant amounts of pollutants, as can the resuspension of settled materials in artificial channels and sewer systems.

Other opportunities for urban runoff pollution are plentiful. Examples include

- Leaking sanitary sewers
- Direct connections of sanitary sewers to storm sewers
- Poorly operating septic systems
- Illegal disposal of oils, paints, etc. to storm sewer systems
- Accidental spills
- Leaking underground storage tanks
- Leachate from landfills
- Leakage from hazardous waste sites

The diversity of urban runoff pollution sources is impressive. Complicating the matter further is the fact that each pollutant comes from a different array of sources. Table 1 links a number of pollutants and their most likely sources. (This table is only indicative of general trends and is not necessarily appropriate for a particular local problem.) In addition to the great variety of sources, pollutants are carried in runoff at varying concentrations. Table 2 presents concentration ranges for some specific pollutants (or indicators of pollution) in urban runoff and, for comparative purposes, other wastewater streams. It is difficult to draw too many generalities from these data. Roughly, urban stormwater is characterized by a heavy suspended solids load when compared to untreated sanitary sewage. The oxygen-demanding potential (as measured by biochemical oxygen demand, BOD) and nutrient levels (as indicated by total nitrogen and total phosphorus) found in urban runoff are somewhat less. Combined sewage, being a combination of sanitary sewage and stormwater runoff, has characteristics somewhere between the two.

Urban runoff contributes to a number of receiving water problems that are as varied as the sources and nature of the pollutants it carries. Table 3 summarizes the impacts associated with various kinds of pollutants.

## 1.4 IS URBAN RUNOFF POINT-SOURCE OR NONPOINT-SOURCE POLLUTION?

The difference between point-source and nonpoint-source pollution is probably more important from a regulatory perspective than a technical one. Point-

**Table 1. Origins of Urban Runoff Pollutants (from Whipple et al., 1983)**

| Pollutant | Soil Erosion | Vehicles: Wear | Vehicles: Exhaust | Industrial Wastes | Fossil Fuels | Lawn and Garden Chemicals | Animal Wastes |
|---|---|---|---|---|---|---|---|
| Suspended solids | M | M | | | m | | |
| Organic material | M | M | m | | | | M |
| Nutrients | | | | | | | |
| Nitrogen | m | | M | | M | M | M |
| Phosphorus | M | | m | | | M | M |
| Petroleum substances | | M | M | M | | | |
| Micro-organisms | | | | | | | M |
| Heavy metals | | | | | | | |
| Iron | M | | | | | | |
| Manganese | M | | | | | | |
| Zinc | m | M | | m | | m | |
| Lead | | | M | m | | | |
| Copper | | M | | M | | | |

| | | | | | | |
|---|---|---|---|---|---|---|
| Chromium | M | | M | | | |
| Nickel | m | | M | | | |
| Cadmium | m | | M | | | |
| Sulfur | | m | | M | | m |
| Acids | | | | | | |
| Nitric | | M | | M | | |
| Sulfuric | | M | | M | | |
| Pesticides | | | | | M | |

M = Major source; m = minor source.

**Table 2. Characteristics of Urban Wastewater Streams (after Ellis, 1986)**

| Wastewater Type | $BOD_5$ (mg/L) | SS (mg/L) | Total N (mg/L) | Total P (mg/L) | Lead (mg/L) | Total Coliforms (MPN/100 mL) |
|---|---|---|---|---|---|---|
| Stormwater | 10–250[c] | 3–11,000 | 3–10 | 0.2–1.7 | 0.03–3.1 | $10^3$–$10^8$ |
| CSOs[a] | 60–200[c] | 100–1,000 | 3–24 | 1–11 | 0.4 | $10^5$–$10^7$ |
| Untreated sewage | 160[c] | 235 | 35 | 10 | — | $10^7$–$10^9$ |
| POTW[b] Effluent | 20 | 20 | 30 | 10 | — | $10^4$–$10^6$ |

[a] Combined sewer overflows.
[b] Publicly owned treatment works.
[c] Single numbers are means. Others represent a range.

**Table 3. Summary of Urban Runoff Pollution Effects (New York State Department of Environmental Conservation, 1992)**

***Sediment* (sand, silt, clay in colloidal suspension)**

Sediment may destroy fish habitat through blanketing of fish spawning and feeding areas and elimination of certain food organisms; directly affect fish through gill abrasion and fin rot; and reduces sunlight penetration, thereby impairing photosynthesis of aquatic plants. Suspended sediment decreases recreational values, reduces fishery habitat, adds to the mechanical wear of water supply pumps and distribution systems, and adds treatment costs for water supplies. Nutrients and toxic substances attached to sediment particles are transported to waterbodies and may enter aquatic food chains, cause fish toxicity problems, impair recreational uses, or degrade the water as a drinking water source.

***Nutrients* (phosphorus, nitrogen)**

Nutrient enrichment of surface waters may cause excessive algae and aquatic plant growth, choking open waters consuming oxygen (mainly through plant die-off); fish and aquatic organisms, fishing and boating, and the use of the resources for water supply are thereby affected. Nitrogen contaminants in drinking water significantly above the drinking water standard may cause methoglobinemia (a blood disease) in infants, and have forced closure of several water supplies.

***Oxygen-Demanding Organics* (human, animal excreta; decaying plant, animal matter; discarded litter, food wastes)**

Organic materials (natural or synthetic) may enter surface waters dissolved or suspended in runoff; natural decomposition of these materials may deplete dissolved oxygen supplies in the surface waters. Dissolved oxygen (DO) may be reduced to below the threshold necessary to maintain aquatic life, harming or killing fish and other aquatic plants and animals

***Toxic Substances* (heavy metals, pesticides, oil, and other petroleum products)**

Toxic chemicals may enter surface waters either dissolved in runoff or attached to sediment or organic materials, and may enter ground waters through soil infiltration. The principal concerns in surface water are their entry into the food chain, bioaccumulation, toxic effects on fish, wildlife, and microorganisms, habitat degradation, and potential degradation of public water supply sources; the ground water impacts are primarily related to water supply sources.

***Pathogens* (bacteria, viruses)**

Bacteria and viruses include infectious agents and disease-producing organisms normally associated with human and animal wastes. The principal concerns are the survival and transmission of such organisms and their impacts on drinking water supplies, contact recreation waters, and fish and wildlife or domestic animals.

**Table 4. Point-Source or Nonpoint-Source Characteristics (from New York State Department of Environmental Conservation, 1990)**

| Point-Source Pollution | Nonpoint-Source Pollution |
|---|---|
| • Pollutants discharged from a single source at a discrete point | • Pollutants entering water at many locations from many sources, distributed diffusely over an area |
| • Pollution can feasibly be abated and/or controlled through regulatory permits, inspections, monitoring, and compliance processes voluntary compliance, or regulation | • Usually best prevented or remediated by modifying activities, practices, or operations on the land, or by changing land use activities financial incentives, either through the use of financial incentives, voluntary compliance, or regulation |
| • Usually controlled through use of wastewater treatment technologies to remove pollutants before discharge | • Usually controlled by reducing or preventing availability, release, or transport of pollutants that adversely affect water quality |
| • Usually associated with the use and disposal of water for industrial, commercial, or municipal purposes | • Usually associated with runoff from precipitation events or with movement of ground water |

source pollution can be defined as pollution emanating from a concentrated location, e.g., a pipe discharging to a stream. Nonpoint-source pollution is that which is not point-source pollution. This is a facetious sounding definition, but it may be the most appropriate. Table 4 lists the general characteristics of point and nonpoint pollution sources as described by one state regulatory agency.

Urban stormwater runoff is, in fact, a *diffuse* source of pollution (Novotny, 1991). In this sense it can be called a "nonpoint" source of pollution. However, in most urban areas runoff is collected in gutters and sewers and discharged at various points to receiving waters (to meet the drainage needs discussed earlier), thus giving it some of the characteristics of a "point" source of pollution. Therefore, is urban runoff point-source or nonpoint-source pollution? Does it really matter? The term diffuse pollution seems better suited.

This book will not embroil itself in what constitutes point-source or nonpoint-source pollution. What primarily characterizes urban runoff, whether eventually discharged from a point or not, is that it is directly linked to precipitation. This, of course, means that urban runoff pollution is not only diffuse, but also episodic, which leads to the correct conclusion that it is relatively difficult to analyze and abate. Regardless of how stormwater is discharged, the pollution it carries represents a threat to many of the nation's waterways.

## 1.5 MODELING AND THE SIMULATION EXPERIMENT

Urban runoff modeling is the *raison d'être* for this book. This section addresses some general concepts that will be useful throughout.

A *model* is a representation of a system in some form other than the system itself (Shannon, 1975). The system in this case is the urban watershed. *Mathematical models*, as the name implies, use mathematical relationships to represent, or simulate, the behavior of the system. It is important to remember that a mathematical model may be as simple as one equation or be comprised of hundreds or thousands of relationships. The engineer who avoids the use of a computer model, claiming that the he or she would not rely on a "model", and then turns around and uses the "rational method", for example, to calculate peak flow from an urban watershed probably misunderstands what the rational method is — a model!

Models help us study systems in ways that would not be practical, or even possible, by other means. *Simulation* is the process of conducting experiments with a model for the purpose of understanding the system or evaluating strategies for the operation of the system (Shannon, 1975). A well-designed model can be used to conduct "simulation experiments" and address any number of "what if" questions in an efficient manner. Computers, with their ability to perform many thousands of computations per second, are ideally suited to the demands of a mathematical model capable of tracking runoff and pollutants through the complex maze that is the urban watershed. Most urban stormwater models (but not all) are computerized.

### 1.5.1 The Simulation Experiment

An urban stormwater model simulates the movement of stormwater and transported materials through an urban watershed in response to precipitation and watershed conditions. A computer model can be a valuable tool to predict an urban watershed's behavior under a variety of conditions (i.e., different design events, different antecedent conditions, changing land uses, etc.). In simulating the watershed's behavior, a model can answer questions that would be impractical to answer through field measurements and observations. This is the essential value of a computer model — it serves as an electronic laboratory. The experiments conducted in this laboratory are not unlike those carried out in any laboratory.

### 1.5.2 Limitations of Simulation

The use of a mathematical model to conduct simulation experiments on an urban watershed is limited by the model itself (just as the experimental setup in a laboratory limits the kind of experiments that can be performed). Every model, whether designed and built by the person using it or acquired from another source, contains the biases and limitations of its creator(s). Even the most sophisticated models are "wrong" to some degree in their representation of the urban watershed and the runoff pollution process. Newcomers to modeling are often frustrated by these limitations. They come of age as modelers when they realize that any model is simply a tool, an imperfect one at that, and they are responsible for the end product. The model user must always interpret the results of a simulation experiment with a keen awareness of model limitations and assumptions, just as a scientist in the laboratory must be aware of the limitations of his or her equipment and methods. In short, applying the same rigor to simulation experiments as one would to laboratory experiments will improve the chances for a successful modeling project.

In addition to the limitations discussed above, these two additional constraints should be heeded:

1. An urban runoff model cannot improve the data base with which it is working. A model can extract information from a data base, but cannot overcome data inadequacies.
2. Output produced by a computer model is no more accurate than numbers produced by hand calculations, just faster. Placing a mathematical model on a computer does not improve that model. The computer model just makes the large number of calculations tractable.

## 1.6 SUMMARY

- Urban stormwater runoff causes both quantity and quality problems in receiving waters. The urban watershed is complex and the pollutant sources, as well as the pollutants, are numerous and extremely diverse.
- The question of whether urban runoff is a "point" source or "nonpoint" source of pollution is not important for the purposes of this book. That it is driven by precipitation is the important, distinguishing feature.
- Mathematical models, computer models in particular, are very useful tools for analyzing urban runoff pollution problems. Models give users the ability to perform experiments on a watershed that would be impractical otherwise. They have limitations, however, and the user must be aware of these and interpret all model results in the light of his or her own knowledge and judgment.

## REFERENCES

Crawford, N.H. and R.K. Linsley, Jr., Digital Simulation in Hydrology: Stanford Watershed Model, Department of Civil Engineering, Stanford University, Technical Report No. 39, July 1966.

Ellis, J.B., "Pollutional Aspects of Urban Runoff," in *Urban Runoff Pollution*, Torno, H., J. Marsalek, and M. Desbordes, Eds., NATO ASI Series, Series G: Ecological Sciences, Vol. 10, Springer-Verlag, Berlin, 1986, p. 1.

Metcalf and Eddy, Inc., University of Florida and Water Resources Engineers, Inc., *Storm Water Management Model, Volume I*, EPA Reports 11024 DOC 10/71, U.S. Environmental Protection Agency, Washington, DC, October 1971.

Moffa, P., S. Nix, P.L. Freedman, J.K. Marr, and R. Field, *Control and Treatment of Combined-Sewer Overflows*, Van Nostrand Reinhold, New York, 1990.

New York State Department of Environmental Conservation, *Nonpoint Source Management Program*, January 1990. New York State Department of Enviromental Conservation, Albany, NY.

New York State Department of Environmental Conservation, *Reducing the Impacts of Stormwater Runoff from New Development,* April 1992. New York State Department of Enviromental Conservation, Albany, NY.

Novotny, V., "Urban Diffuse Pollution: Sources and Abatement," *Water Environ. Technol.*, Vol. 3, No. 12, December 1991, pp. 60–65.

Shannon, R.E., *Systems Simulation: The Art and Science*, Prentice-Hall, Englewood Cliffs, NJ, 1975.

U.S. Environmental Protection Agency, *Results of the Nationwide Urban Runoff Program, Volume I, Final Report*, NTIS PB84-185552, U.S. Environmental Protection Agency, Washington, DC, 1983.

U.S. Environmental Protection Agency, *National Water Quality Inventory: 1986 Report to Congress*, U.S. Environmental Protection Agency, Washington, DC, 1986.

Federal Register, NPDES Permit Application Regulation for Stormwater Discharges, 40CFR, part 122–124, *55FR477990*, November 16, 1990.

Walesh, S.G., *Urban Surface Water Management*, John Wiley & Sons, New York, 1989.

Whipple, W., S. Grigg, T. Grizzard, C.W. Randall, R.P. Shubinski, and L.S. Tucker, *Stormwater Management in Urbanizing Areas*, Prentice-Hall, Englewood Cliffs, NJ, 1983.

# 2 URBAN STORMWATER MODELS

## 2.1 OVERVIEW

There are a number of well-respected urban stormwater models available to the water resources and environmental engineering communities. The discussion here assumes that the reader is interested in selecting one of these as opposed to constructing one of his or her own design. While there are merits in building your own model, most notably that you have a model that meets your needs *and* you understand how it works, can be very time consuming and probably not very cost effective. In addition, such a model will not undergo the scrutiny experienced by the well-established models.

This chapter will introduce basic modeling concepts, explore the kinds of models available, and discuss some of these models in detail. Chapter 3 will provide guidance on which model to select.

## 2.2 GENERAL MODELING TERMINOLOGY

A model has been defined as a representation of a system. A mathematical model uses mathematical relationships to represent the system. Beyond those simple definitions, models can be placed somewhere on a continuum between two extremes: deterministic and stochastic models.

Strictly speaking, a *deterministic* (or *mechanistic*) *mathematical model* is a model that has everything right. In other words, the model is a perfect reflection of the physical world it is trying to simulate. Therefore, it should be able to

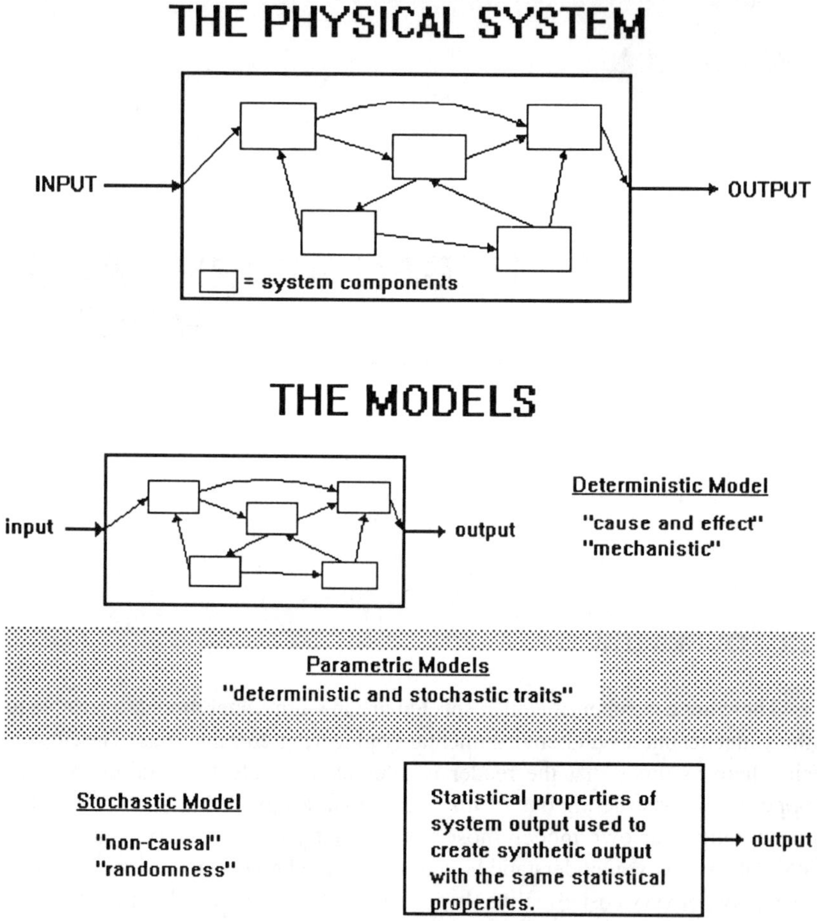


**Figure 1. Deterministic vs. stochastic models.**

predict exactly what will happen, for example, when a rainstorm moves over an urban watershed. (If this were well understood we would probably also be able to predict precisely when and where the rainstorm will occur with some other model.) Such a model would always produce the same answer every time it encountered the same set of inputs, because everything is accurately depicted. On the other hand, a *stochastic mathematical model*, in the strictest sense again, eschews understanding the cause-effect relationships of the physical world in favor of using the statistical patterns of a particular phenomenon to simualte that same phenomenon. A stochastic model will produce different sets of output from the same input, although the output will have fairly consistent statistical properties. The differences between the two types of models are illustrated in Figure 1.

Of course, only the extreme cases are discussed above. No model is a perfect representation of the world, nor does any model completely ignore causal relationships. Essentially all models are *parametric models* that include deterministic and stochastic traits. Many models have the deterministic trait of always producing the same output from a given set of inputs. However, appearances are deceiving. Consider the simple model:

$$R = k \cdot P \tag{2.1}$$

where $R$ = runoff, inches
$P$ = precipitation, inches
$k$ = coefficient (or parameter)

If $k = 0.6$, you will always get 1.2 inches of runoff every time $P$ is 2 inches. However, the coefficient $k$ is loaded with uncertainty. Giving it a number such as 0.6 does not mean that it is certain, nor is the model itself certain; in fact, it is probably inadequate for anything but crude estimates. The model parameter is a constant in the context of the model, but it is likely that $k$ has considerable variability.

A model that produces sequences of monthly streamflow through the use of the streamflow's statistical properties, such as its probability distribution, has the stochastic trait of producing different sets of streamflows for a given set of inputs (in this case the moments of the distribution). However, it may also use the prediction for month N to influence the value predicted for month N+1...a deterministic trait! Perhaps it may use the properties of the watershed to estimate the parameters of the probability distribution. Again, this is a deterministic trait.

Most urban runoff models are parametric models that feel deterministic; perhaps they should be called "pseudo-deterministic". They have varying amounts of uncertainty or randomness embedded in their general formulation, input parameters, and input data, but a particular set of parameters and input data will always produce the same output. For the purposes of this book, models that produce identical outputs for a given set of inputs will be called deterministic. Those that produce different outputs for a given set of inputs will be called stochastic. This is an imperfect but fairly common use of the terms.

In recent years, model developers and users have begun to deal more openly with the uncertain or stochastic side of urban runoff models; this will be discussed in later chapters. Recognizing uncertainty has caused model results to be viewed more critically. This has been a most welcome development and enhances the use of models to study urban runoff pollution problems.

Models may also be classified by the way in which spatial differences are handled. Watershed characteristics change from one point to another. Deterministic models (now in the loose sense of the term) that account for spatial

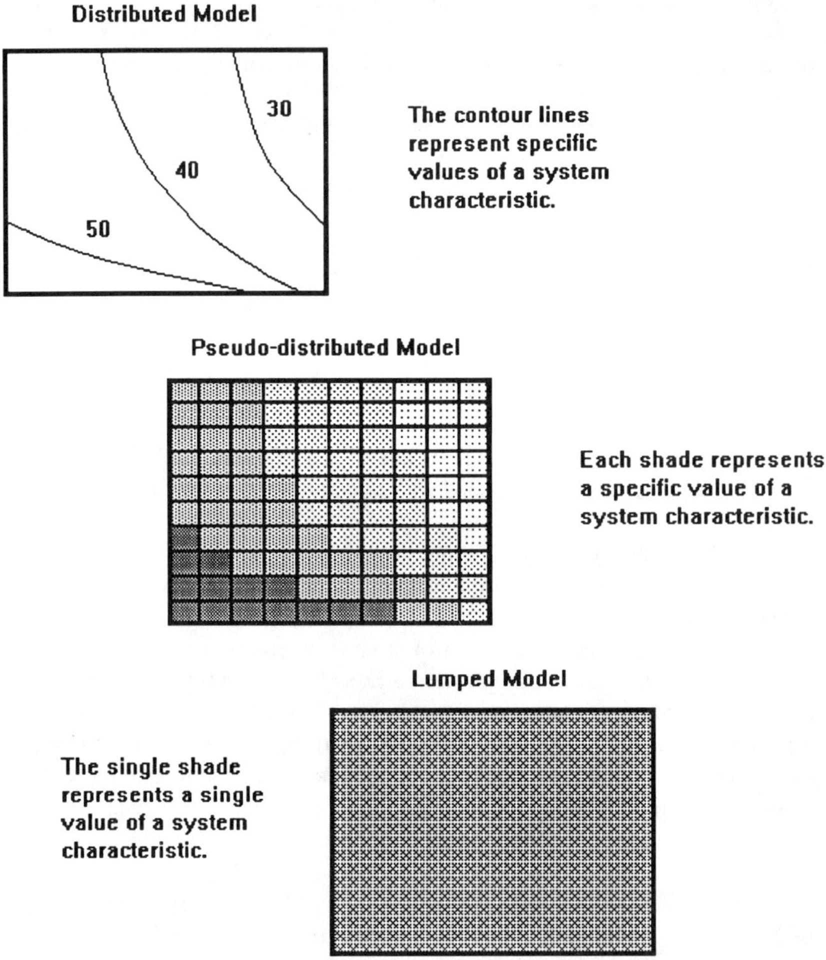


**Figure 2. Distributed vs. lumped models.**

variability as a function of the position in the watershed are known as *distributed models* (see Figure 2). Models that assume that all characteristics are constant over the watershed are known as *lumped models*. A lumped model can be used to create a "pseudo-distributed" model if the watershed is divided into a number of subwatersheds. Each subwatershed has uniform characteristics within its boundaries, but when they are grouped to form the entire watershed some measure of spatial variability is created. Most runoff models use this approach.

The vast majority of urban runoff models are "pseudo-deterministic", "pseudo-distributed" models. The focus here will be on these models and their application to urban runoff problems.

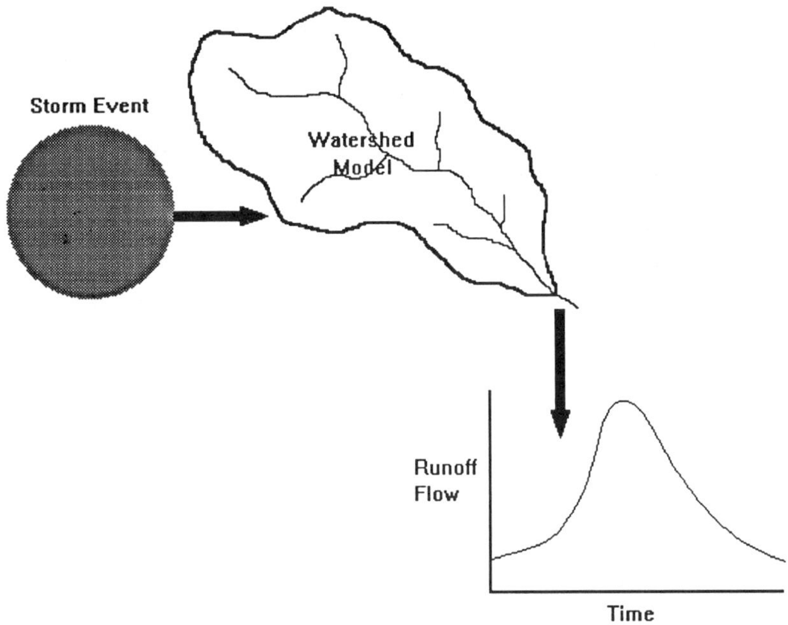


**Figure 3. Single-event simulation.**

Models are also classified in terms of the time frame to which they are applied. *Single-event models* simulate the behavior of a watershed over a single rainfall-runoff event (see Figure 3). *Continuous models* are designed to simulate watershed behavior over many sequential events. They are distinguished from single-event models in that they must be able to simulate the hydrologic changes that occur in a watershed between storms (see Figure 4). Because of the computational time necessary to simulate the hydrologic processes over long periods, continuous models tend to model the watershed in less detail than single-event models. The benefits and problems associated with both types are discussed in Chapter 3.

Each urban runoff model is an imperfect representation of the watershed and its ability to simulate the watershed adequately must be validated. *Validation* refers to the process of collecting data to describe the inputs and outputs to an urban watershed for a wide range of conditions *and* adjusting the model parameters so that the model adequately replicates the watershed as depicted by those data. Validation is vital to the creation of a credible model and it will be described in Chapter 3, along with some other tools.

The following modeling and simulation definitions will also be useful in this chapter and the rest of the book. A *run* is a single execution of a model. *Run time* is the time needed to execute a model run on a computer. *Simulation time* is the time simulated by a model run.

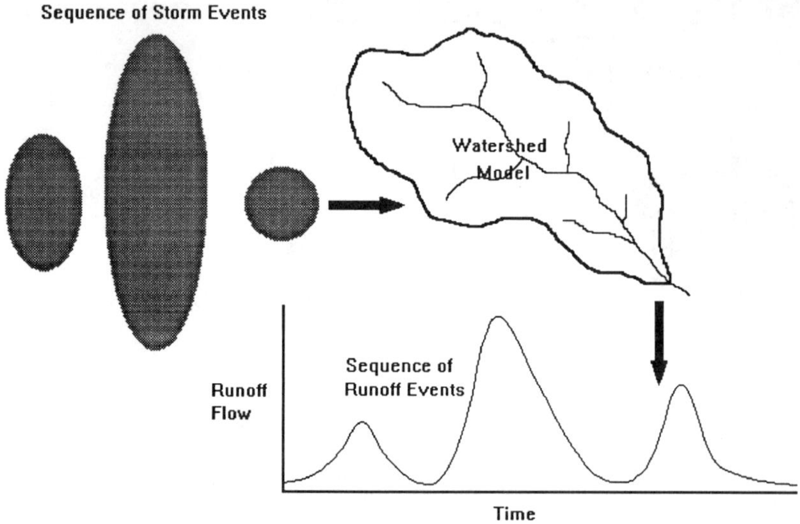


**Figure 4. Continuous simulation.**

## 2.3 SOME USEFUL HYDROLOGIC AND WATER QUALITY TERMINOLOGY

The hydrologic and water quality terms defined in this section are presented in the interest of completeness and consistency. To most readers they will be familiar, but a review is useful.

*Rainfall depth* is the volume of rainfall occurring over a specified area expressed as an equivalent depth over that area (e.g., inches or millimeters). (The term precipitation depth is perhaps more appropriate, but the terms rainfall and precipitation will be used synonymously throughout the book.) *Rainfall intensity* is the rate, in terms of depth per unit time (e.g., inches per hour or millimeters per hour), at which rainfall occurs over an area. A *hyetograph* is a plot of rainfall intensity vs. time (see Figure 5). Hyetographs are almost always displayed as average intensities over finite intervals of time (e.g., hourly intensities, inches per hour or millimeters per hour, over 5-minute intervals) so that a typical one looks like a bar chart as opposed to the smooth event it actually is. This is known as a quantized-continuous series of data (see Chapter 3).

The quantity of runoff is expressed as a *runoff volume* (e.g., cubic feet, acre-feet, or cubic meters) or *runoff depth*, which is the volume of runoff discharged from a specified area expressed as an equivalent depth over that area (e.g., inches or millimeters). The *runoff flow rate*, or simply the *runoff rate*, is the volume or depth per unit of time (cubic feet per second, cubic meters per second, or acre-feet per hour for volume over time; inches per hour or millimeters per hour for depth over time) at which runoff occurs from an area

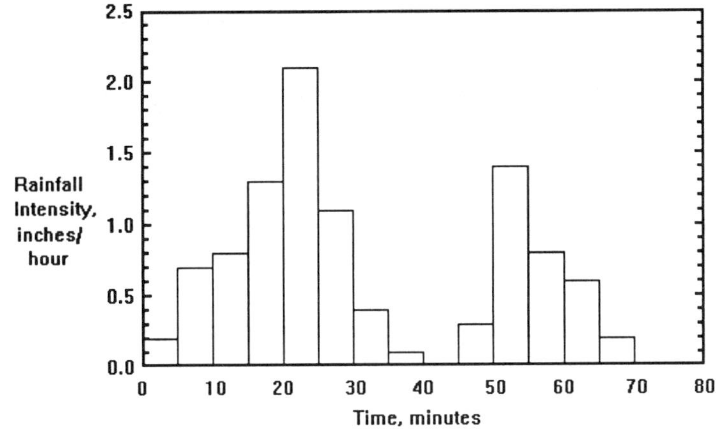


**Figure 5. Precipitation hyetograph.**

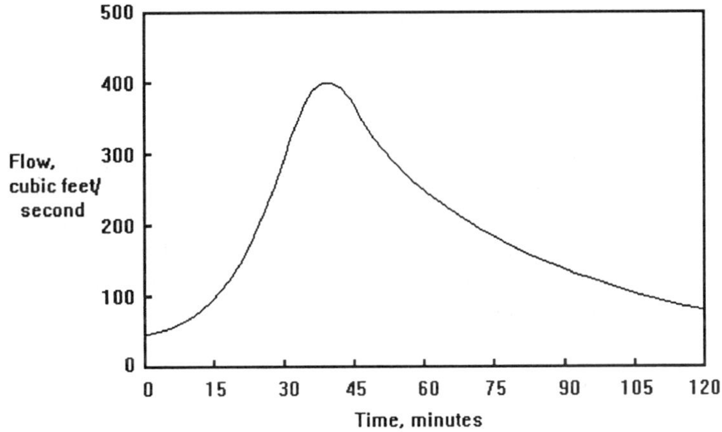


**Figure 6. Runoff hydrograph.**

(usually, but not necessarily, a *watershed,* the area contributing runoff to a point). A *hydrograph* is a plot of the runoff flow rate vs. time (see Figure 6).

The quantity of a pollutant discharged from an area is known as the *pollutant load* and is expressed in terms of mass or mass per unit area (e.g., pounds or kilograms; pounds per acre or kilograms per hectare). The *pollutant load rate* is the load discharged per unit time (e.g., pounds per day or kilograms per day; pounds per acre-day or kilograms per hectare-day). The ratio of pollutant mass (or occasionally another measure of quantity such as number of coliform bacteria) to volume is, of course, *pollutant concentration* (e.g., pounds per cubic foot or milligrams per liter). A *pollutograph* is a plot of the pollutant concentration (or sometimes the pollutant load rate) vs. time (see Figure 7).

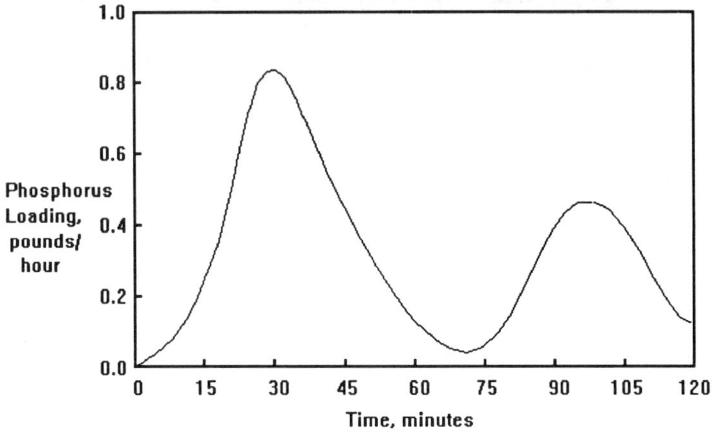


**Figure 7. Runoff pollutograph.**

## 2.4 GENERAL MODEL STRUCTURE

Conceptually, most models are rather simple. Most of the complexity resides with "bookkeeping" and input/output functions. This is important to keep in mind. Urban runoff models use the same basic hydrology, hydraulics, and pollutant transport concepts that most users learned in school. The watersheds to which they are applied are complex and, thus, there is a large amount of accounting needed to track water and pollutants within and between the various processes. In addition, considerable effort is often expended in producing impressive looking output or "user friendly" input procedures. Models are immensely useful, but feel free to be unimpressed. Strive to understand what they do and how they do it and keep a critical eye.

From a general perspective, a model is made up of the components shown in Figure 8. The "core" of any urban runoff model is, of course, the mathematical relationships used to simulate the behavior of the urban watershed and the procedures used to solve them. These relationships are comprised of input, output, and state variables as well as coefficients or parameters.

*Input variables* are quantities that change over time, influence system (i.e., watershed) behavior, and enter the system, or model of the system, without being affected by other system variables, i.e., they are independent. (They may not be truly independent, but, from the standpoint of the system or model of the system and its boundaries, they are.) *State variables* are quantities within the system that change over time in response to input variables and other state variables. *Output variables* are quantities that change over time in response to the input and state variables and leave the system.

*Parameters*, or *coefficients*, are quantities that do not change with time, that is, they do not change over the course of the simulation experiment. However, one must remember that the difference between a variable and a parameter is

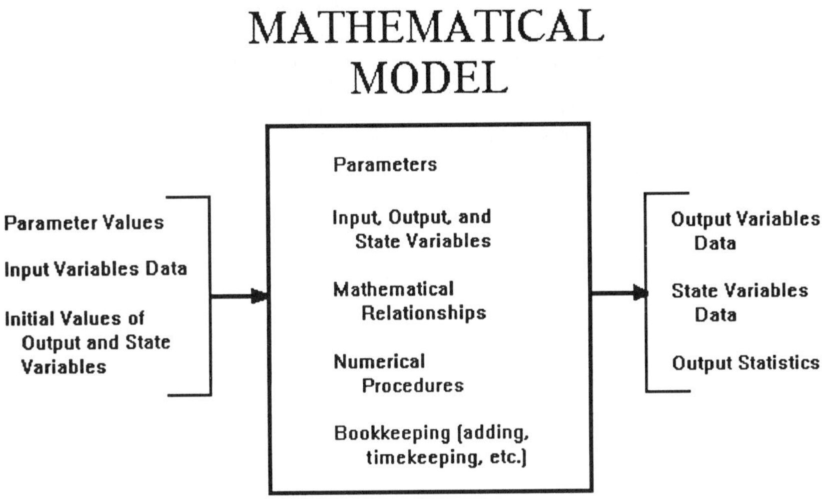


**Figure 8. Model components.**

really just a matter of time. A parameter, such as the percent of the urban watershed covered by impervious surfaces, may be static in the short term, but will probably change over several years.

Numerical procedures are important parts of most models, since they are often needed to solve the mathematical relationships for the values of the state and output variables (these are discussed in more detail later in this chapter). In addition, there are a number of algorithms to sum quantities (e.g., pollutant loads from the watershed), keep time, and perform statistical analyses.

From a computer programming standpoint, most urban runoff models have the general structure shown in Figure 9. The "traffic director" coordinates the flow of the program and moves data from one algorithm to another. Input processors read the input variable data (e.g., rainfall intensities), initial values of state variables (e.g., soil moisture), and parameters. The computational algorithms carry out the important calculations, much of their work involved in carrying out the numerical procedures used to solve the mathematical representations of the system. Output processors display the output variables (e.g., runoff flow rate over time) and state variables. They may also perform various statistical analyses on these data and report the results.

## 2.5 URBAN STORMWATER MODELS

As noted earlier, there are many models in the marketplace capable of simulating urban runoff quantity and quality to one degree or another. Probably the best way to classify these models is to look at how they route flow and pollutants through the urban watershed (Moffa et al., 1990; Nix, 1991). In other

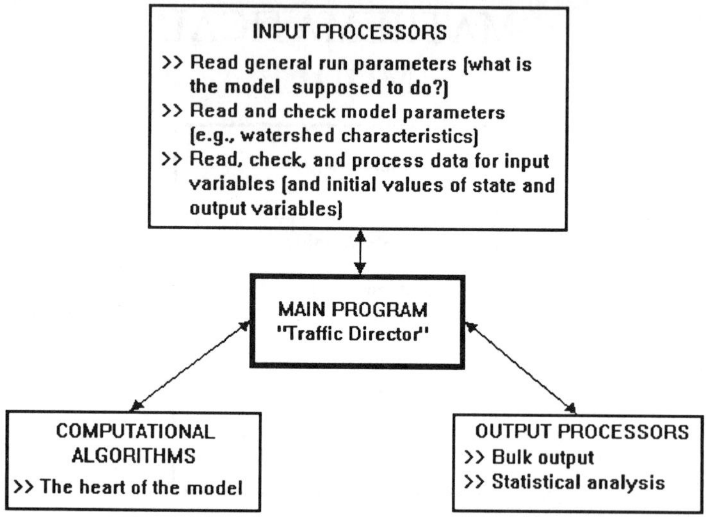


**Figure 9. Program components.**

words, how much detail and sophistication does the model use to track water and pollutants? Three levels of detail and, thus, three categories of models, are discussed here. The features of the three levels are shown in Figure 10 and discussed below.

At the first level are very simple models that do not route flows and pollutants through the watershed, require little input data, and may need little, if any, help from a computer. Their principal purpose is to calculate long-term averages (e.g., annual runoff volumes and pollutant loads). These estimates are useful for locating problem areas and sometimes suggesting an abatement strategy. Examples include SWMM Level I (Heaney et al., 1976) and the statistically based procedures developed by Howard (1976), DiToro and Small (1979), and Hydroscience, Inc. (1979).

At the second level of sophistication are models that are computer executed, generate runoff hydrographs (flow vs. time) and pollutographs (concentration vs. time) at one point, or at relatively few points, in a watershed, and allow some simple routing of flow and pollutants through channels and storage basins. These are usually designed to simulate watershed behavior over long periods (i.e., continuous simulation). This feature allows the user to estimate the characteristics of many runoff events over long planning periods. Examples include the Corps of Engineers Storage, Treatment, Overflow, Runoff Model or STORM (Hydrologic Engineering Center, 1977) and the Hydrological Simulation Program — Fortran or HSPF (Johanson et al., 1984).

At the third, and last, level are models capable of routing flows through gutters, channels, and sewers and producing runoff hydrographs and

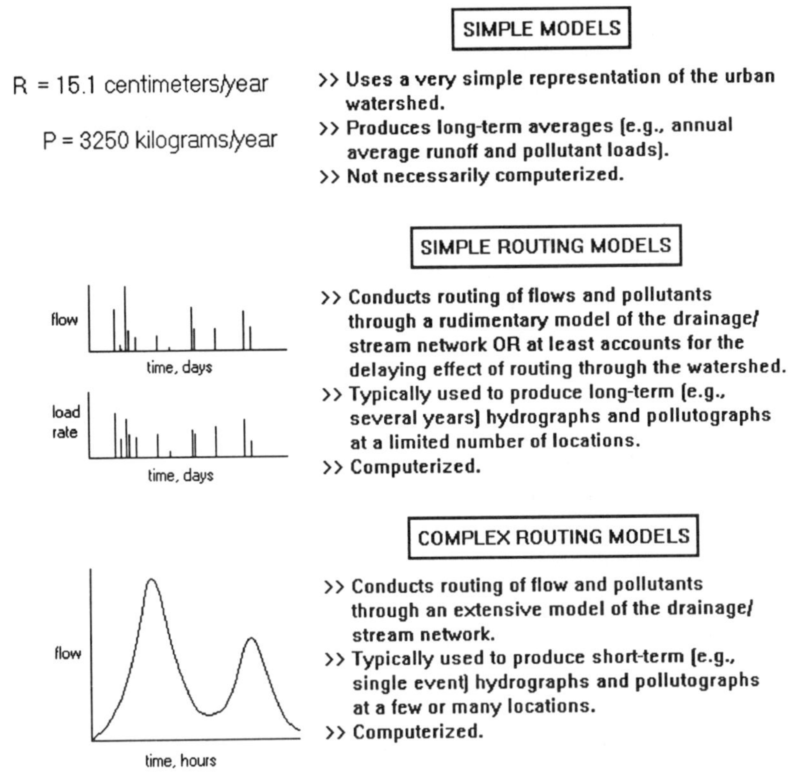


**Figure 10. Levels of urban runoff models.**

pollutographs at various locations in the urban watershed. Some are intended for single-event simulation and others are capable of continuous simulation. Some of the more well-known models of this type include the EPA Storm Water Management Model or SWMM (Huber and Dickinson, 1988; Roesner et al., 1988) and the Quantity-Quality Simulator or QQS (Geiger and Dorsch, 1980).

This three-tiered classification scheme is useful because the detail to which the urban watershed is modeled says much about the objectives of the modeling effort. If we are only interested in event volumes and loads from a watershed, a model that routes flows and pollutants throughout the watershed may be unnecessary. However, if we are interested in how those event flows and pollutant loads behave over time and/or at many points in the watershed, a model capable of more sophisticated routing may be in order. The reader should keep in mind, however, that some of the most sophisticated models are capable of performing simpler analyses.

The different levels of models can be used in a hierarchical attack on a problem. This is done by using the results produced by a simpler model (or simpler version of a more complex model) to guide the use of a more sophisticated model. In fact, it is possible that a sequence of several models or a single model applied at various degrees of detail and complexity could be used to serve many needs.

A number of publications have summarized and compared models (American Public Works Association, 1981; Brandstetter, 1976; Chu and Bowers, 1977; Hall, 1984; Huber, 1986a, 1986b; Huber and Heaney, 1980, 1982; U.S. Environmental Protection Agency, 1983; Whipple et al., 1983). The reader should refer to these publications for more detailed discussions of the available models.

Table 1 summarizes the models mentioned above. The next several sections will discuss, in some detail, three models representative of each of the levels of complexity. No endorsements of these particular models are intended, but all are good examples and can give the reader an understanding of the range of available capabilities. The reader should not be confused by the detail of the presentations. The discussions are not intended to be comprehensive treatises on the models, but rather they are meant to clearly explain their basic premises so that they may be used more effectively and confidently. Each model has a user's manual that can (and should) be consulted for detailed explanations of all model functions. The reader may find the discussion here to be more useful than that found in the various user's manuals. The author certainly hopes this is the case. Nevertheless, the primary intent is to piece common modeling techniques into a more cohesive picture.

Before looking at the three representative models, one further note on water quality terminology is in order. It is fairly common to refer to quality *indicators* like biochemical oxygen demand (BOD) and coliform bacteria as "pollutants". While this is incorrect, no attempt will be made here to correct the usage. For convenience, all water quality constituents will be called "pollutants".

Some of the units used in the following descriptions will be different from those used in the model documents. Changes are introduced to maintain clarity and consistency in *this* document. However, some symbols appear in more than one context. They are clearly redefined, but be aware.

## 2.6 STORM WATER MANAGEMENT MODEL: LEVEL I

The Storm Water Management Model: Level I (SWMM Level I) was developed to meet the need for a simple, easy-to-apply tool capable of providing preliminary assessments of urban stormwater problems and potential control strategies (Heaney et al., 1976). Its name is unfortunate, because it bears little or no resemblance to the far more complex Storm Water Management Model (SWMM, the last model discussed in this chapter, see Section 2.8). It is not a computer-based model. It was, however, intended to be the "bottom-

**Table 1. Characteristics of Representative Urban Runoff Models (adapted from Nix, 1991)**

| Model | Routing Level | Time Domain | Predictive Method | Pollutant Transport |
|---|---|---|---|---|
| SWMM — Level I | None | Long-term averages | Empirical | None |
| Statistical | None | Long-term averages | Frequency distribution | None |
| STORM | Simple storage | Continuous and single event | Buildup and washoff; soil loss | Mass balance |
| HSPF | Simple storage channel and routing | Continuous and single event | Buildup and washoff; soil loss | Completely mixed |
| QQS | Channel, storage, and pipe routing | Continuous and single event | Buildup and washoff | Plug flow |
| SWMM | Channel and pipe routing | Continuous and single event | Buildup and washoff; soil loss; rating curve | Completely mixed; plug flow (in storage) |

of-the-line" model in a series of models leading up to the full-fledged version of SWMM. This series was never fully developed, but SWMM Level I is a useful tool for estimating annual runoff totals and pollutant loads.

### 2.6.1 Quantity Estimation

SWMM Level I is designed to estimate the average annual runoff from a watershed with a minimum of input data. The model is based on a very simple relationship:

$$AR = C \cdot P \tag{2.2}$$

where $AR$ = average annual runoff, inches per year
$C$ = runoff coefficient
$P$ = average annual precipitation, inches per year

Equation 2.2 is modified to account for losses due to depression storage (the precipitation held in surface depressions and foliage and not appearing as runoff) with a relationship taken from the Storage, Treatment, Overflow, Runoff Model (STORM, the second model discussed in this chapter, see Section 2.7):

$$DS = 0.25 \cdot (1 - I/100) - 0.0625 \cdot (I/100) \tag{2.3}$$

where $DS$ = weighted maximum depression storage, inches
$I$ = percent of the watershed that is impervious

The constants 0.25 and 0.0625 represent the assumed maximum depression storages, in inches, for pervious and impervious areas, respectively.

To estimate the effect of depression storage on annual runoff, one year of precipitation data for Minneapolis were used in a number of STORM simulation runs, each with a different value of depression storage (actually different values of $I$). The impact on annual runoff was found to fit a power equation, i.e.,

$$\text{“annual effect” (in inches)} = 5.234 \cdot (DS)^{0.5957} \tag{2.4}$$

This result modifies Equation 2.2 as follows:

$$AR = (C \cdot P) - 5.234 \cdot (DS)^{0.5957} \tag{2.5}$$

SWMM Level I adapts another relationship used by STORM to estimate the runoff coefficient (see Section 2.7):

$$C = 0.15 \cdot (1 - I/100) + 0.90 \cdot (I/100) \tag{2.6}$$

The constants 0.15 and 0.90 are the assumed runoff coefficients for pervious and impervious areas, respectively. The value of $C$ is weighted between these two constants depending on the imperviousness of the watershed.

Imperviousness, indicated by $I$, is estimated with a relationship developed by Stankowski (1974):

$$I = 9.6 \cdot PD_d^{\,0.573 - 0.0391 \cdot \log_{10} PD_d} \tag{2.7}$$

where $PD_d$ = population density, persons per acre

This equation was developed using data collected from several hundred watersheds in New Jersey.

A review of Equations 2.2 through 2.7 reveals that the average annual runoff is estimated as a function of precipitation and population density. This creates a very simple model needing only two pieces of information.

### 2.6.2 Quality Estimation

An analysis of stormwater quality data by Heaney et al. (1976, 1977) led to the following pollutant loading equations:

$$ms_{i,j} = \alpha_{i,j} \cdot P \cdot f_i(PD_d) \cdot \delta_i \tag{2.8}$$

or

$$mc_{i,j} = \beta_{i,j} \cdot P \cdot f_i(PD_d) \cdot \delta_i \tag{2.9}$$

where
$ms_{i,j}$ = average annual load of pollutant $j$ from land use $i$ within an area served by separate storm sewers, pounds per acre-year
$mc_{i,j}$ = average annual load of pollutant $j$ from land use $i$ within an area served by combined sewers, pounds per acre-year
$\alpha_{i,j}$ = loading factor for pollutant $j$ and land use $i$ within areas served by separate sewers, pounds per acre-inch
$\beta_{i,j}$ = loading factor for pollutant $j$ and land use $i$ within areas served by combined sewers, pounds per acre-inch
$P$ = average annual precipitation, inches per year
$f_{i(}PD_{d)}$ = population density function for land use $i$
$\tau_i$ = street sweeping effectiveness factor for land use $i$

Equations 2.8 and 2.9 make a distinction between the pollutant loads emanating from areas with separate sewers and those served by combined sewers. The average annual load of pollutant $j$ from the entire urban area can be calculated as

$$M_j = P \cdot \sum_{i=1}^{n} \left[ \left( ms_{i,j} \cdot AS_i \right) + \left( mc_{i,j} \cdot AC_i \right) \right] \tag{2.10}$$

where
$M_j$ = overall annual load of pollutant $j$ from the urban area, pounds per year
$AS_i$ = area of land use $i$ within areas served by separate sewers, acres
$AC_i$ = area of land use $i$ within areas served by combined sewers, acres
$n$ = number of land uses (SWMM Level I assumes four land uses)

Table 2 presents the means to calculate the various components of Equations 2.8 and 2.9.

### 2.6.3 Preliminary Cost Assessment for Pollution Control

A very interesting and useful feature of SWMM Level I is its ability to provide preliminary estimates for the costs of controlling stormwater pollution.

**Table 2. SWMM Level I Pollutant Loading Functions (from Heaney et al., 1976)**

The following equations may be used to predict annual average loading rates as a funtion of land use, precipitation and population density.

$$ms_{i,j} = \alpha_{i,j} \cdot P \cdot f_i(PD_d) \cdot \gamma_i$$

$$mc_{i,j} = \beta_{i,j} \cdot P \cdot f_i(PD_d) \cdot \gamma_i$$

where $M$ = pounds of pollutant $j$ generated per acre of land use $i$ per year,
$P$ = annual precipitation, inches per year,
$PD_d$ = developed population density, persons per acre,
$\alpha,\beta$ = factors given in table below,
$\gamma$ = street sweeping effectiveness factor, and
$f_i(PD_d)$ = population density function.

*Land Uses:*
$i$ = 1, residential
$i$ = 2, commercial
$i$ = 3, industrial
$i$ = 4, other developed, e.g., parks, cemeteries, schools (assume $PD_d = 0$)

*Pollutants:*
$j$ = 1, $BOD_5$, total
$j$ = 2, suspended solids (SS)
$j$ = 3, volatile solids, total (VS)
$j$ = 4, total $PO_4$ (as $PO_4$)
$j$ = 5, total N

*Population Function:*
$i$ = 1 $f_i(PD_d) = 0.142 + 0.218 \cdot PD_d^{0.54}$
$i$ = 2,3, $f_i(PD_d) = 1.0$
$i$ = 4 $f_i(PD_d) = 0.142$

*Factors $\alpha$ and $\beta$ for Equations:* Separate factors, $\alpha$, and combined factors, $\beta$, have units lb/acre-in. To convert to kg/ha-cm, multiply by 0.442.

| | | Pollutant, *j* | | | | |
|---|---|---|---|---|---|---|
| | Land Use, *i* | 1. $BOD_5$ | 2. SS | 3. VS | 4. $PO_4$ | 5. N |
| Separate areas, $\alpha$ | 1. Residential | 0.799 | 16.3 | 9.45 | 0.0336 | 0.131 |
| | 2. Commercial | 3.20 | 22.2 | 14.0 | 0.0757 | 0.296 |
| | 3. Industrial | 1.21 | 29.1 | 14.3 | 0.0705 | 0.277 |
| | 4. Other | 0.113 | 2.70 | 2.6 | 0.00994 | 0.0605 |

| | | | | | | |
|---|---|---|---|---|---|---|
| | 1. Residential | 3.29 | 67.2 | 38.9 | 0.139 | 0.540 |
| Combined | 2. Commercial | 13.2 | 91.8 | 57.9 | 0.312 | 1.22 |
| areas, β | 3. Industrial | 5.00 | 120.0 | 59.2 | 0.291 | 1.14 |
| | 4. Other | 0.467 | 11.1 | 10.8 | 0.0411 | 0.250 |

*Street Sweeping:* Factor $\gamma_i$ is a function of street sweeping interval, $N_s$, (days):

$$\delta_i = \begin{Bmatrix} N_s/20 & \text{if } 0 \leq N_s \leq 20 \text{ days} \\ 1.0 & \text{if } N_s > 20 \text{ days} \end{Bmatrix}$$

---

The assumed control mechanism is the storage/treatment configuration shown in Figure 11. This configuration is the same as that used by STORM (see Section 2.7). This configuration is assumed to work as follows. Runoff occurring at a rate less than the treatment capacity is passed through the "treatment" facility and released. Runoff exceeding the capacity of the treatment facility is sent to a storage basin. When the storage basin reaches its capacity the excess runoff is routed to the receiving water. Runoff held in storage is returned to the treatment plant when the runoff rate is less than the treatment capacity.

Simulations performed with STORM were used to develop performance functions (or what are also known as "production" functions) for the storage/treatment facility hypothesized in Figure 11 for five cities around the U.S.: Washington, DC, Atlanta, San Francisco, Denver, and Minneapolis. These cities were selected to represent the climates in five regions. These regions are shown in Figure 12. A "typical" year of hourly rainfall data was selected from 25-year records and used to generate runoff flows and pollutant loads from each city. These flows and loads were routed through various storage capacity-treatment rate combinations to create a performance function for pollutant control for each city (BOD was used):

$$R = f(S,T) \qquad \textbf{(2.11)}$$

where
$R$ = long-term percent pollutant (BOD) control
$S$ = storage capacity, inches
$T$ = treatment capacity, inches per hour

The units used in Equation 2.11 are normalized over the watershed area (i.e., the volume of water represented by 1 inch of runoff is 1 inch × watershed area). The term "control" refers to the amount of pollutant that passes through treatment (STORM does not simulate removal in treatment). Pollutant "removal" is estimated as the product of R and an assumed treatment efficiency.

The performance function developed for Minneapolis is shown in Figure 13. Each line in this figure is known as an isoquant and represents a particular pollutant control percentage. The performance function for each of the five key

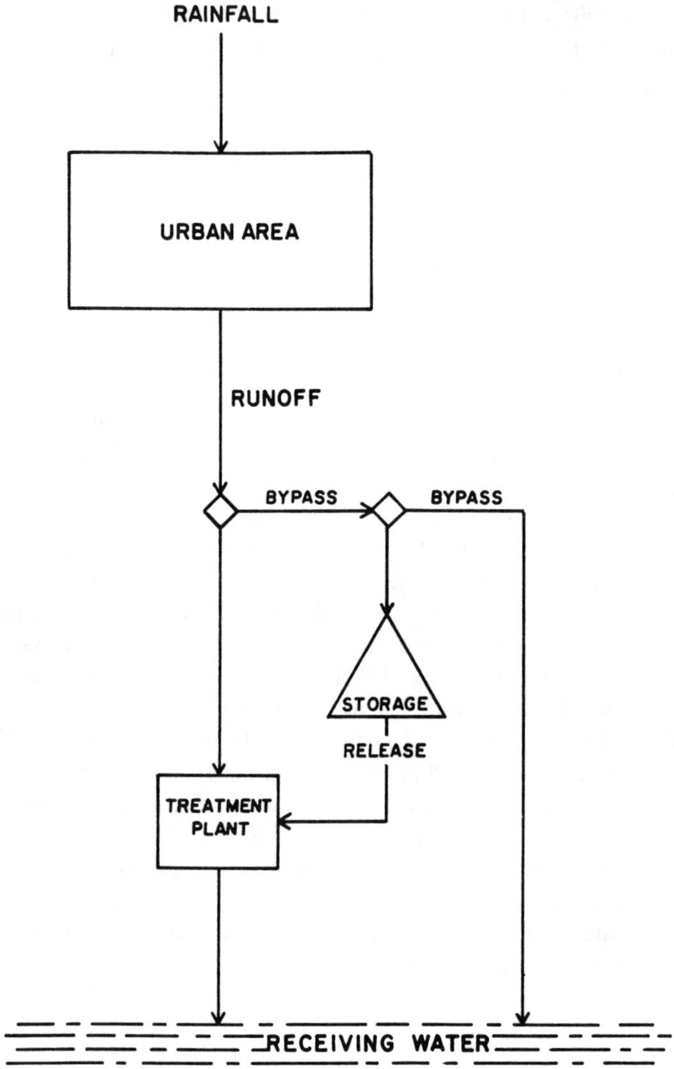


**Figure 11. Storage/treatment configuration, SWMM Level I (after Heaney et al., 1976).**

cities can be modified for other cities in its region (based on the differences in annual runoff as calculated by Equation 2.5).

Constant unit costs for treatment and storage are used to create the cost function:

$$Z = c_S \cdot S + c_T \cdot T \tag{2.12}$$

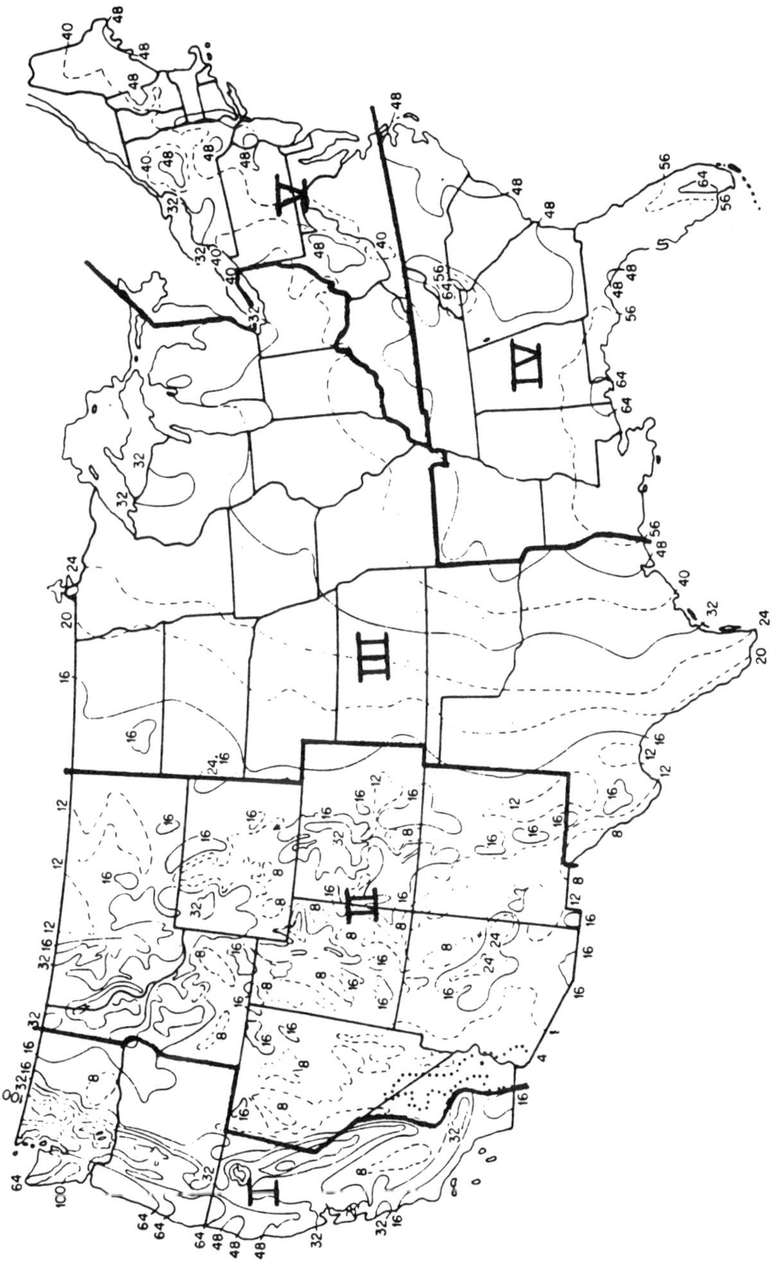

Figure 12. Five climatic regions, SWMM Level I (from Heaney et al., 1976).

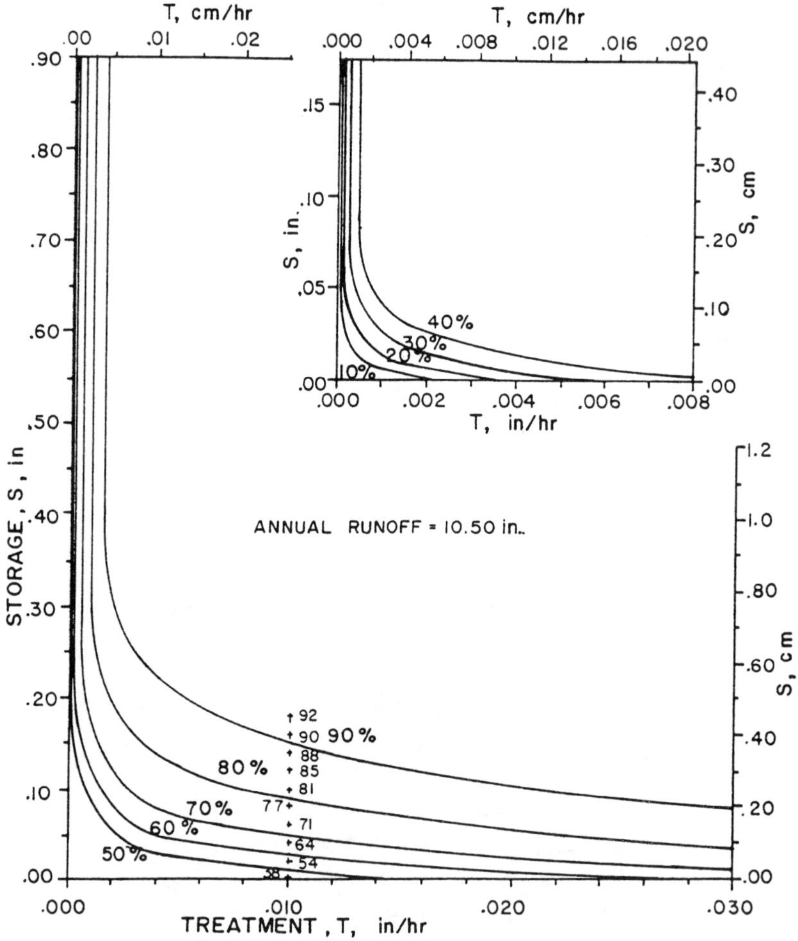


**Figure 13. Storage/treatment performance curves for BOD control, Minneapolis, SWMM Level I (from Heaney et al., 1976).**

where $Z$ = storage/treatment cost, dollars per acre
$c_S$ = unit cost of storage, dollars per acre-inch
$c_T$ = unit cost of treatment, dollars per acre-inch per hour

All costs are expressed as annual costs. Equations 2.11 and 2.12 are then used to construct the following optimization problem:

$$\text{Minimize } Z = c_S \cdot S + c_T \cdot T \qquad \textbf{(2.13)}$$

$$\text{such that } R^\circ = f(S,T) \qquad \textbf{(2.14)}$$

where $R^\circ$ = specific long-term pollutant control percentage

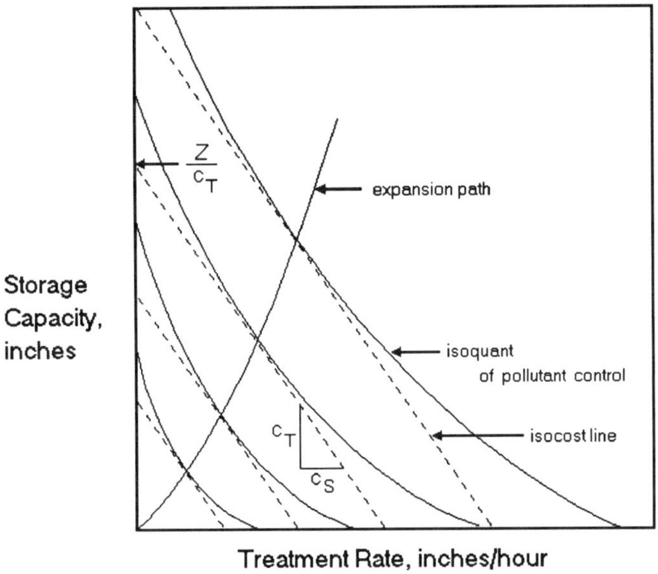


**Figure 14. Locating the optimal storage/treatment strategy, SWMM Level I (after Heaney et al., 1976).**

An exponential function was fit to the relationship represented by Equation 2.14 so that the problem could be solved analytically. The details of this method are found in the SWMM Level I documentation. For the purposes of illustration, however, Figure 14 will suffice. For any one level of pollutant control, $R°$, there is a least-cost combination of S and T. This least-cost combination occurs where the line representing Equation 2.13, known as the isocost line, first comes in contact with the isoquant as $Z$ is increasing. Connecting these locations creates a line known as an expansion path. The expansion path represents the least-cost combinations of storage and treatment to achieve various levels of pollutant removal. In other words, as pollution control expands, it does so along this line (if done optimally). Subsequently, a relationship between cost and pollutant control can be developed:

$$Z^* = g(R) \tag{2.15}$$

where $Z^*$ = least cost to achieve pollutant control percentage $R$

A typical cost relationship is shown in Figure 15.

### 2.6.4 Typical SWMM Level I Application

During the first round of concern about nonpoint source pollution, i.e., the 1970s, the Larimer-Weld Regional Council of Governments prepared a plan

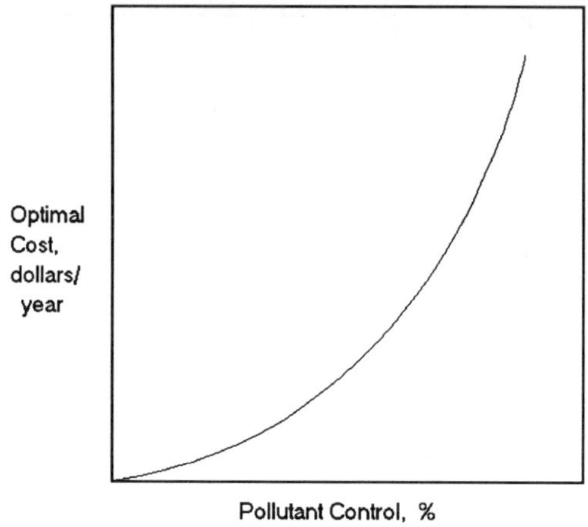


**Figure 15. Final storage/treatment cost as a function of BOD control, SWMM Level I (after Heaney et al., 1976).**

for nonpoint pollution control as part of a larger water quality management plan (Pitts et al., 1977). This area is north of Denver and includes the counties of Larimer and Weld and the cities of Fort Collins, Loveland, and Greeley (see Figures 16 and 17).

Urban stormwater loads in the Larimer-Weld region for the years 1975 and 2000 were calculated with SWMM Level I. The results are summarized in Table 3. After a review of the receiving waters and their assimilative capacities and the contribution from other sources, nitrogen and BOD loads from urban runoff were deemed to be relatively insignificant. On the other hand, urban suspended solids loads were felt to be a far more significant source than the solids discharged from sanitary sewage treatment plants in the region. The study went on to develop preliminary control costs for storage/treatment using the SWMM Level I methodology summarized above, as well as a companion methodology to evaluate and compare structural and nonstructural control technologies (Heaney and Nix, 1977). The costs for storage/treatment (as a function of BOD removal) in Fort Collins are shown in Figure 18.

SWMM Level I played an important part in this study and helped the Larimer-Weld Regional Council of Governments move forward and intelligently plan their next move. This is the proper role for simple models such as SWMM Level I.

### 2.6.5 Summary

SWMM Level I is a simple model intended to give the user preliminary estimates of average annual runoff and pollutant loads, as well as control costs. It has not been implemented for the use on a computer; it is applied with tables,

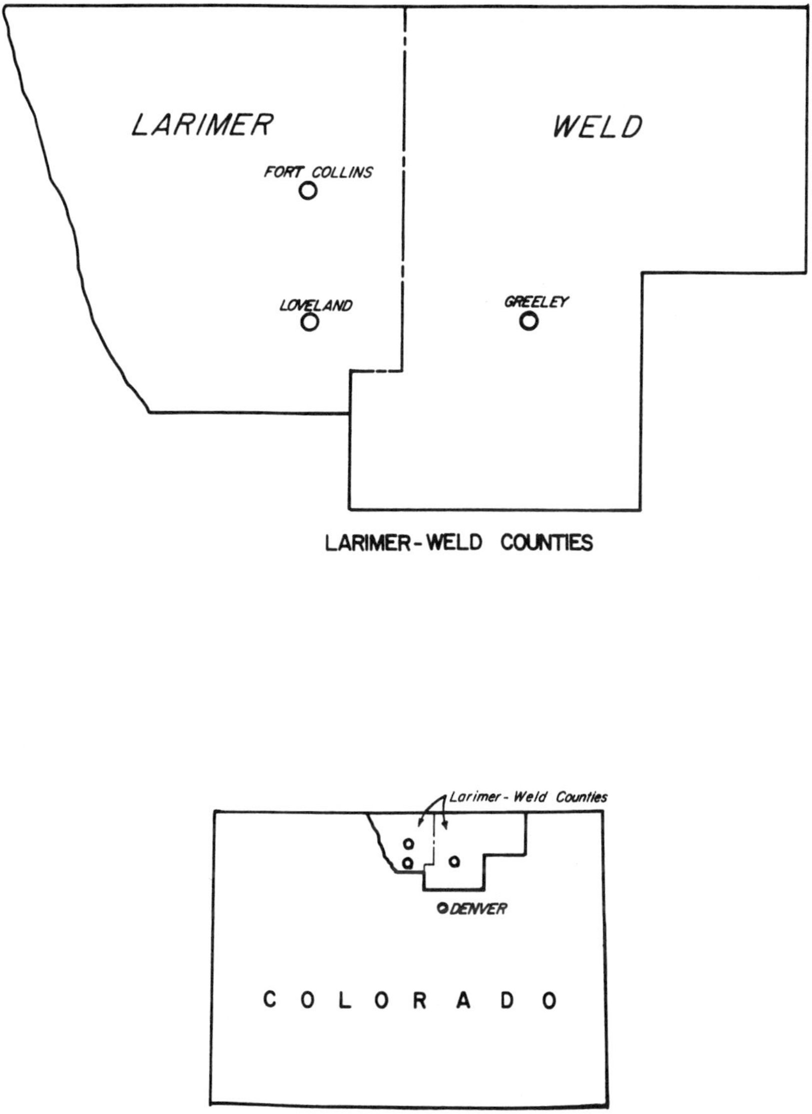


**Figure 16. Larimer-Weld region (from Pitts et al., 1977).**

nomographs, and a minimum of input data. It is best suited for quick assessments of *relative* load contributions and control costs.

## 2.7 CORPS OF ENGINEERS STORAGE, TREATMENT, OVERFLOW, RUNOFF MODEL

The Corps of Engineers' Storage, Treatment, Overflow, Runoff Model (STORM) is a software package originally released in 1973. Significant modi-

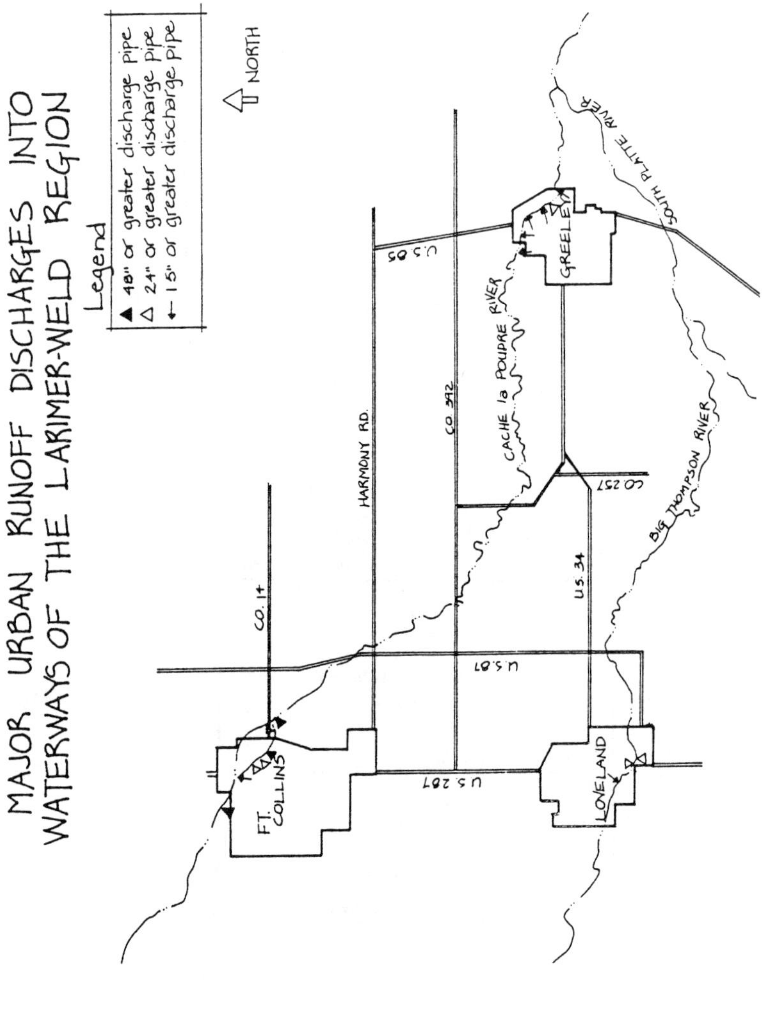


**Figure 17. Major urban runoff discharges into waterways of the Larimer-Weld region (from Pitts et al., 1977).**

Table 3. Average Annual Urban Runoff Pollutant Loads for the Larimer-Weld Region (from Pitts et al., 1977)

| | Municipality | | | | | |
|---|---|---|---|---|---|---|
| | Fort Collins | | Greeley | | Loveland | |
| Year | 1976 | 2000 | 1976 | 2000 | 1976 | 2000 |
| Acreage | 11,700 | 26,600 | 13,134 | 19,000 | 6,000 | 10,900 |
| Population density (people/acre) | 4.8 | 5.6 | 4.4 | 6.8 | 4.2 | 5.6 |
| Annual stormwater (acre-ft/year)[a] | 3,900 | 10,100 | 1,800 | 3,560 | 2,075 | 4,140 |
| Suspended solids | | | | | | |
| (lbs/acre-year)[b] | 155 | 165 | 123 | 148 | 146 | 165 |
| (tons/year)[c] | 1,013 | 2,410 | 913 | 1,510 | 498 | 990 |
| $BOD_5$ | | | | | | |
| (lbs/acre-year)[b] | 7.6 | 8.1 | 6.0 | 7.3 | 7.1 | 8.1 |
| (tons/year)[c] | 57 | 136 | 51 | 85 | 28 | 56 |
| Nitrogen | | | | | | |
| (lbs/acre-year)[b] | 1.2 | 1.3 | 1.0 | 1.2 | 1.2 | 1.3 |
| (tons/year)[c] | 8.7 | 21 | 8 | 13 | 4.6 | 8.5 |

[a] Based on average rainfall developed from 25 years of record.
[b] Residential land use only.
[c] All land uses.

fications continued until 1977 when the final release appeared (Hydrologic Engineering Center, 1977). In more recent years, the Corps' Hydrologic Engineering Center has turned the task of distribution over to several vendors. Information and copies of the STORM software can be obtained from Dodson Associates, 6529 FM 1960 West, Suite 314, Houston, Texas 77069, 800-235-8069 (voice) or 713-440-4742 (fax) or Haestad Methods, 37 Brookside Road, Waterbury, Connecticut 06708, 800-727-6555 (voice) or 203-597-1488 (fax). Information about other HEC models and modeling activities (which is extensive) can be obtained by writing to the U.S. Army Corps of Engineers, Hydrologic Engineering Center, 609 Second Street, Davis, California or contacting the Center at 916-756-1104 (voice) or 916-756-8250 (fax).

In spite of its age, STORM remains a viable and commonly used model. It is capable of simulating the runoff and pollutant discharges from urban and rural watersheds in response to precipitation occurring over any length of time (i.e., single event and continuous simulations are easily conducted). Loads and

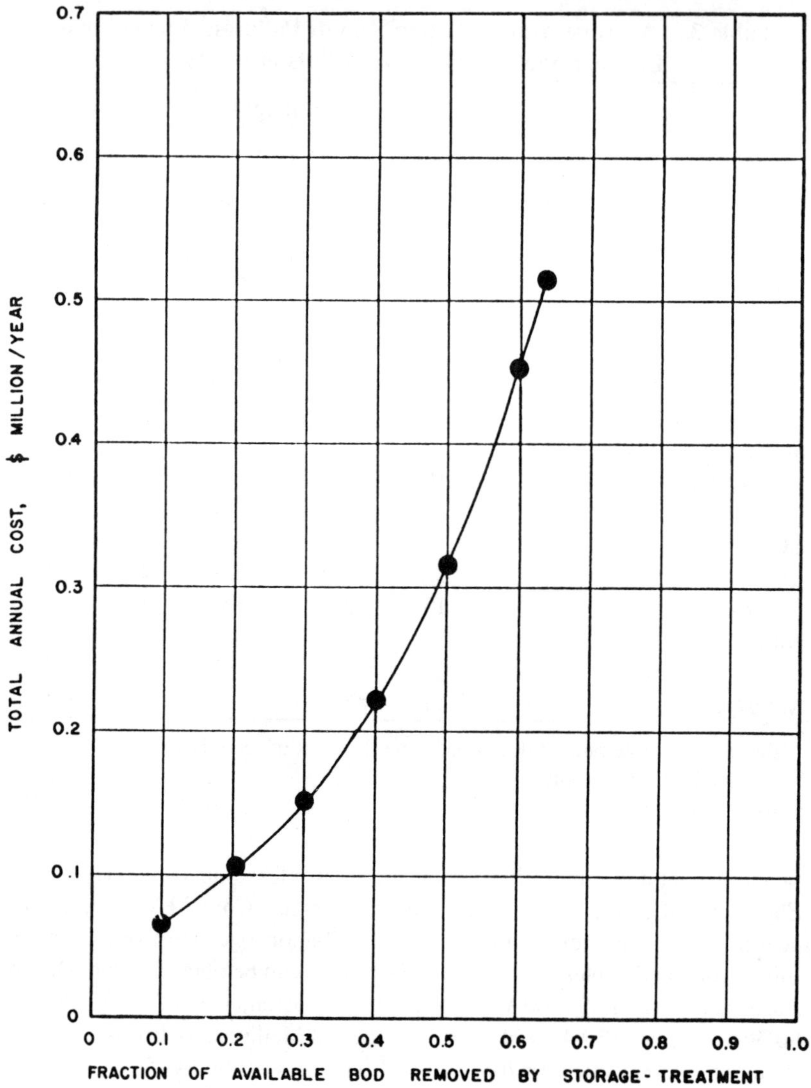


**Figure 18. Storage/treatment cost as a function of BOD removal, Fort Collins, CO (from Pitts et al., 1977).**

concentrations for six water quality parameters are computed, including suspended solids, settleable solids, biochemical oxygen demand, total nitrogen, orthophosphate, and total coliforms. Soil erosion is also modeled.

STORM's strengths are its simplicity and its routing capability (although it is a rudimentary capability). STORM's simplicity allows it to simulate watershed response over many years in very short order. This, in turn, allows analysis of the statistical nature of the runoff and load events.

The basic components modeled in STORM are shown in Figure 19. A portion of the rainfall becomes runoff which washes pollutants off the watershed. The runoff and pollutants are then routed to a storage/treatment facility. The various components are discussed below.

STORM is written in FORTRAN. It is not particularly user friendly in the modern sense of the term, but it simplicity makes its management and use straightforward.

### 2.7.1 Quantity Simulation

The principal input variable to STORM is hourly precipitation. A watershed may be divided into a number of subwatersheds; however, total runoff is simply the summation of the results from the subwatersheds and no attempt is made to route the flows through the watershed. (STORM does have, as mentioned earlier, simple routing capability, but it resides mostly in the storage/treatment algorithms. This will be discussed later.)

Hourly runoff is calculated for each subwatershed in one of three ways: the coefficient method, the Soil-Complex-Cover method, and the unit hydrograph. The *coefficient method* is the most commonly used and calculates runoff during each hour as

$$R = C \cdot (P - f) \tag{2.16}$$

where
$R$ = runoff rate, inches/hour or millimeters per hour;
$C$ = composite runoff coefficient
$P$ = precipitation intensity, inches per hour or millimeters per hour
$f$ = available depression storage, inches or millimeters
$DS$ = maximum depression storage, inches or millimeters

The composite runoff coefficient is calculated as a function of land use:

$$C = C_P + (C_i - C_p)\sum_{i=1}^{n} X_i \cdot F_i \tag{2.17}$$

where
$C_P$ = runoff coefficient for pervious surfaces
$C_I$ = runoff coefficient for impervious surfaces
$X_i$ = fraction of land use i in subwatershed

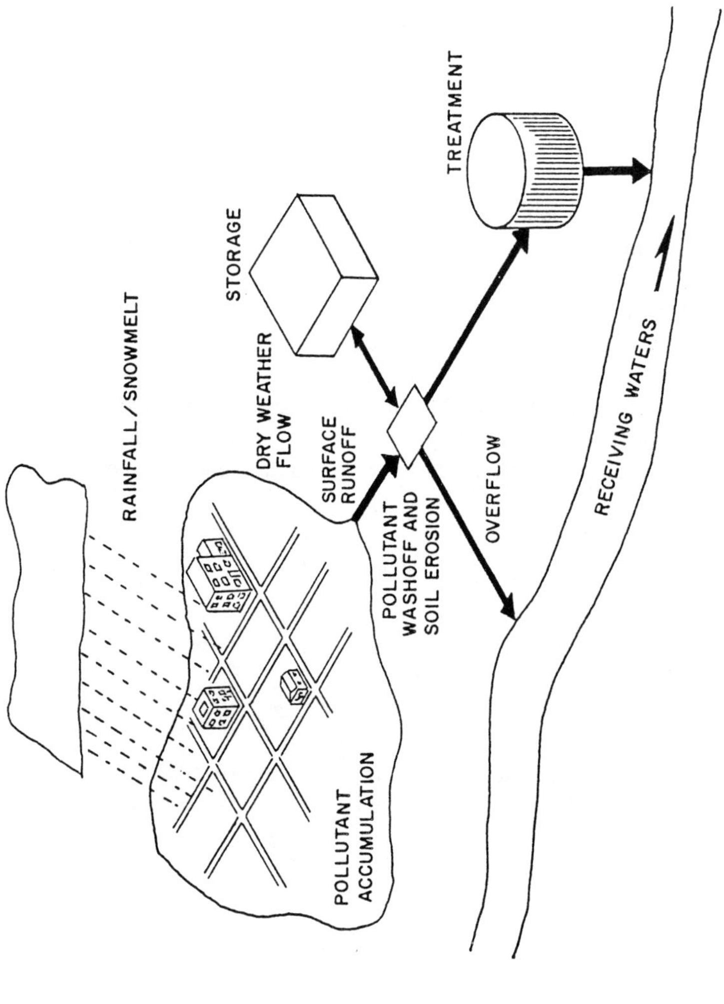


**Figure 19. Major processes modeled by the Storage, Treatment, Overflow, Runoff Model, STORM (from Hydrologic Engineering Center, 1977).**

$F_i$ = fraction of land use $i$ that is impervious
$n$ = number of land uses

(This equation is identical to Equation 2.6 in the description of SWMM Level I if the summation term, which is the fraction of the subwatershed that is impervious, is replaced by I/100, and if $C_P$ and $C_I$ are assumed to be 0.15 and 0.90, respectively.)

As discussed previously, the maximum depression storage, DS, is the *capacity* of the subwatershed to store water in depressions and on foliage. This capacity is reduced as precipitation occurs and is replenished by evaporation. The available depression storage at any point in time, which must be less than or equal to $DS$, is calculated as

$$f = f_o + N_D \cdot k_e, f \leq DS \tag{2.18}$$

where $f_o$ = available depression storage after previous rainfall, inches or millimeters
$N_D$ = elapsed time from the previous storm, days
$k_e$ = recovery rate of depression storage, which is assumed to be the pan evaporation rate, inches per day

As Equation 2.16 shows, runoff does not occur unless the precipitation for any given hour exceeds the available depression storage, $f$. Figure 20 summarizes the way the coefficient method estimates hourly runoff. Figure 20 also shows how treatment, storage, and overflows are simulated (these features will be discussed later in this section).

There are two other methods for calculating runoff. The second method relies on the venerable *Soil-Complex-Cover Method* developed by the Soil Conservation Service (1985). The technique is based on an empirically derived relationship between storm rainfall and direct runoff (i.e., surface runoff not stored on the surface or infiltrated into the soil).

The third method of runoff simulation is based on the *unit hydrograph.* A unit hydrograph is a direct runoff hydrograph resulting from 1 unit (1 inch, 1 millimeter, etc.) of excess rainfall (rainfall that is neither retained on the surface nor infiltrated to the soil; in terms of volume it is equal to direct runoff). A unit hydrograph must be specified for a particular watershed and a particular period of excess rainfall. (For example, a 1-hour unit hydrograph is a direct runoff hydrograph resulting from 1 unit of excess rainfall occurring over 1 hour.) A unit hydrograph is used to calculate a runoff hydrograph by scaling the unit hydrograph by the excess rainfall depth. The advantage of the unit hydrograph method is that it accounts for the *effect* of routing through the subwatershed (at least in terms of how the effect is manifested at the subwatershed outlet).

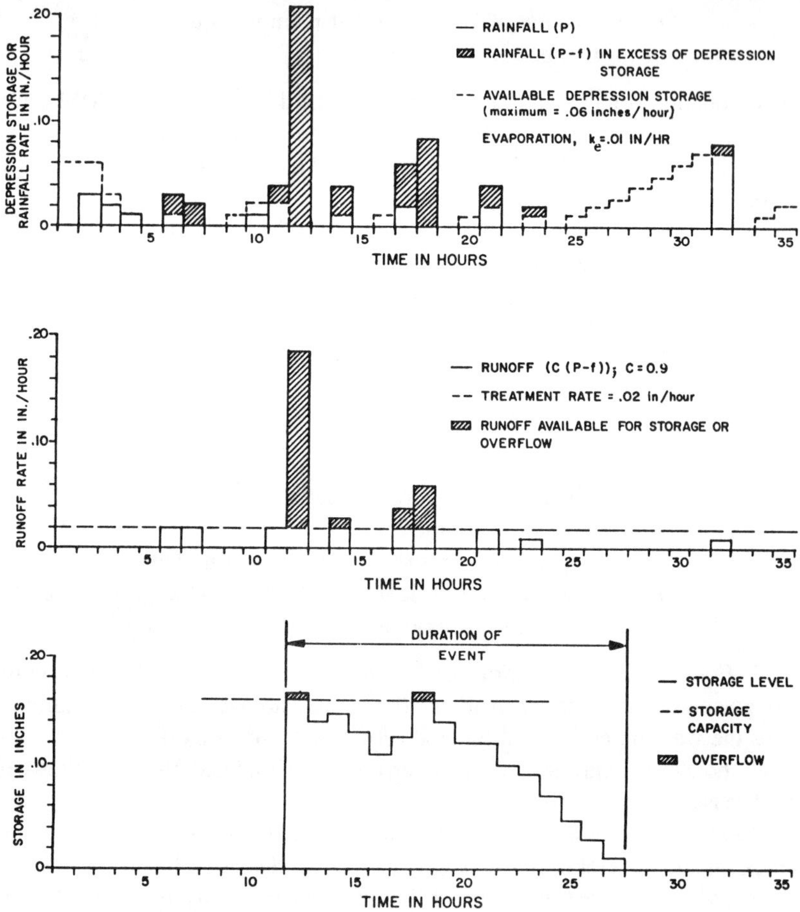


**Figure 20. Histories of rainfall, runoff, storage, and overflows, STORM (from Hydrologic Engineering Center, 1977).**

## 2.7.2 Quality Simulation

STORM simulates the quality of runoff in a fairly straightforward manner. Six pollutants are simulated (suspended solids, settleable solids, and total nitrogen, orthophosphate, biochemical oxygen demand, total coliform bacteria) and each is treated in an essentially identical way.

A pollutant is assumed to accumulate on the land surface in one of two ways. The first assumes that the amount of pollutant is associated with the amount of "dust and dirt" accumulated on streets. This assumption comes out of a well-known study on the debris found on the streets of Chicago (American Public Works Association, 1969). The Chicago study determined the daily accumulation of dust and dirt per foot of street in several areas and the amounts of several pollutants per unit quantity of dust and dirt. Using this study as its foundation, STORM requires the user to specify a dust and dirt accumulation rate for streets in each different land use in the watershed and the amount of each pollutant per unit of dust and dirt.

The following equations (Equations 2.19 through 2.26) are applied to each land use in the watershed. The subscript *i*, which is intended to identify a particular land use, is dropped in Equations 2.20 through 2.26 to avoid excessive clutter.

The rate at which dust and dirt accumulates is given by:

$$DD_i = dd_i \cdot (G_i / 100) \cdot A_i \qquad \textbf{(2.19)}$$

where
- $DD_i$ = rate of dust and dirt accumulation on land use *i*, pounds per day or kilograms per day
- $dd_i$ = rate of dust and dirt accumulation per unit length of street gutter in land use *i*, pounds per day per 100 feet of gutter or kilograms per day per 100 meters of gutter
- $G_i$ = length of gutter per unit of area in land use *i*, feet per acre or meters per hectare
- $A_i$ = area of land use *i*, acres or hectares

The amount of pollutant on each land use before the beginning of a storm is calculated as

$$P_p = (F_p \cdot DD_i \cdot N_D) + P_{po} \qquad \textbf{(2.20)}$$

where
- $F_p$ = amount of pollutant *p* on land use *i* at the beginning of a storm, pounds or kilograms
- $F_p$ = amount of pollutant *p* per unit of dust and dirt, pounds per pound or kilograms per kilogram
- $N_D$ = elapsed time without runoff since the last storm, days
- $P_{po}$ = amount of pollutant *p* on land use *i* at the end of the last storm, pounds or kilograms

The second method of accumulation is much simpler and is based on a rate of accumulation per unit area, thus foregoing the need for "dust and dirt" data on city streets. This simple accumulation method may appear to be intended for nonurban areas and the dust-and-dirt method may seem to be intended for urban areas. In fact, both can be used in either arena since the dust-and-dirt method can be manipulated to be identical to the simple accumulation method by making sure that the product $F_P DD_i$ (in Equation 2.20) is equal to the accumulation, in pounds/day or kilograms/day, from the area associated with a particular land use.

The way in which pollutants are washed off the watershed during a precipitation event is independent of the method used to accumulate the pollutants. STORM assumes that the rate at which a pollutant disappears from (i.e., washes off) land use $i$ in each subwatershed is proportional to the amount remaining (Roesner, 1977):

$$\frac{dP_p}{dt} = -K \cdot P_p \tag{2.21}$$

where $P_p$ = amount of pollutant p on land use $i$ at time $t$, pounds or kilograms
$t$ = time from start of storm, hours
$K$ = coefficient, hours$^{-1}$

Integrating Equation 2.21 between time $t$ and time $t + \Delta t$ later results in

$$P_p(t + \Delta t) = P_p(t) \cdot \exp(-K \cdot \Delta t) \tag{2.22}$$

where $P_p(t + \Delta t)$ = amount of pollutant $p$ on land use $i$ at time t + $\Delta t$, pounds or kilograms
$P_p(t)$ = amount of pollutant $p$ on land use $i$ at time $t$, pounds or kilograms
$\Delta t$ = time interval, one hour

The principal deficiency suffered by Equations 2.21 and 2.22 is that the runoff rate has no effect on the rate of washoff. This problem is addressed by defining $K$ as a function of the runoff rate from impervious areas:

$$K = k \cdot R_I \tag{2.23}$$

where $k$ = washoff decay coefficient, inches$^{-1}$ or millimeters$^{-1}$,
$R_I$ = runoff rate, inches per hour or millimeters per hour

If the coefficient method is used, $R_I$ is assumed to be the runoff rate from impervious areas (since this is where most of the runoff is originating). Substituting for $K$ in Equation 2.22 yields

$$P_p(t + \Delta t) = P_p(t) \cdot \exp\left(-k \cdot R_I \cdot \Delta t\right) \tag{2.24}$$

The rate of removal of mass from land use $i$ over $\Delta t$ is

$$M_p = \left[P_p(t) - P_p(t + \Delta t)\right] / \Delta t = P_p(t) \cdot \left[1 - \exp\left(-k \cdot R_I \cdot \Delta t\right)\right] / \Delta t \tag{2.25}$$

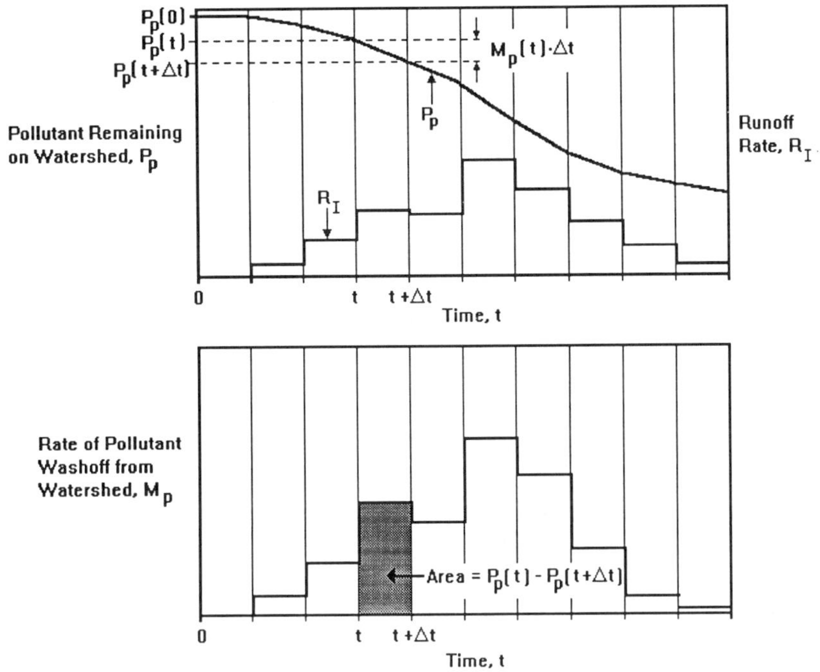


**Figure 21. Development of pollutograph, STORM (after Roesner, 1976).**

where $M_p$ = pollutant $p$ washed off land use $i$ in a given hour, pounds per hour or kilograms per hour

In order to account for the likelihood that not all of the accumulated pollutant will be available for washoff with runoff at any given time, Equation 2.25 is multiplied by an "availability factor":

$$M_p = A_p \cdot P_p(t) \cdot \left[1 - \exp\left(-k \cdot R_I \cdot \Delta t\right)\right] / \Delta t \qquad \textbf{(2.26)}$$

where $A_p$ = availability factor for pollutant $p$

The factor $A_p$ is a different function for each pollutant. Each, however, is a function of $R_I$, the runoff rate from impervious areas. Figure 21 demonstrates how Equations 2.24 and 2.26 work to generate pollutant loads.

STORM also has the capability of simulating dry-weather flows and pollutant loads from residential, commercial, and industrial sources as well as sewer infiltration. Soil erosion is also modeled with an algorithm based on the Universal Soil Loss Equation.

### 2.7.3 Storage/Treatment Simulation

The ability of STORM to simulate a storage/treatment facility is limited, but useful. The routing procedure is elementary. Refer again to Figure 19 and to Figure 20(b) and (c). Runoff occurring at a rate less than the treatment capacity passes through the plant, and is released. If the runoff rate exceeds the capacity of the treatment plant the excess is sent to a storage basin. When the storage basin is full, excess runoff is routed to the receiving water. Runoff held in storage is returned to the treatment plant when the runoff rate is less than the treatment capacity. While the age of water in storage is tracked, there are no provisions for modeling the removal of pollutants in either the treatment plant or storage basin. Pollutants are simply passed through. However, the amount that passes through treatment is tracked.

An *event* is defined as beginning when storage is required and ending when storage is empty. This approach insures that events are defined with respect to the *response* of the urban watershed, including storage and treatment (or storage distributed throughout the watershed). Runoff that just passes through the treatment plant will not register as an event and is essentially treated as an insignificant occurrence. Figure 20 illustrates the effect of the storage/treatment system on a sequence of hourly runoff values and how an event is defined.

The storage/treatment algorithm can be used to simulate, in a very simple way, a number of situations. The following are some example scenarios that could be handled:

- Stormwater detention basin
- Combined sewer interceptor (where the treatment rate is analogous to the interceptor capacity)
- Wet-weather storage/treatment facility
- Storm sewer storage
- Reservoir/lake storage

The fact that STORM is principally intended for continuous simulation makes it very effective for assessing the behavior of "storage/treatment" systems over long periods.

### 2.7.4 Presentation and Analysis of Model Output

One of the features that makes STORM so useful is its presentation and analysis of continuous simulation output. The program produces four output reports: quantity analysis, quality analysis, pollutograph analysis, and land surface erosion analysis. The quantity and quality reports include event-by-event information on runoff, pollutant discharges, and overflows (including pollutants) from storage. These two reports also include annual summaries of rainfall, runoff, pollutant discharges, and the quantity and quality of the storage overflows. The erosion report includes event-by-event information on sediment discharges as well as average annual sediment production and delivery. The pollutograph report produces detailed information about the discharge of pollutants within selected events.

### 2.7.5 Typical Application

STORM is well suited for assessing the long-term behavior of a watershed, including a storage/treatment system. An example of such an application was referred to in Section 2.6 and was shown in Figure 13. This figure presents the results of numerous runs of STORM for the developed areas in Minneapolis. Each run was set up for a different combination of storage capacity, S, and maximum treatment rate, T. The quality report displays how much of the pollutant load (BOD, in this case) passes through the treatment unit. This is referred to as "BOD control" in Figure 13 (inferring that by passing through treatment there was an opportunity for removal). By making several runs the lines of equal control, or isoquants, can be drawn. This kind of information can then be used to develop control strategies with the kind of economic analysis suggested in Figure 14.

### 2.7.6 Summary

The STORM software package remains popular for good reason. It is simple and effective at what it is designed to do, which is to study the long-term behavior of an urban watershed and runoff control devices. STORM simulates runoff and pollutant loads with simple algorithms for any length of the time the user desires (limited only by length of the rainfall record). Flow and pollutants are routed through a simple storage/treatment configuration that can be used to analyze a number of routing and control alternatives. The output is designed to provide useful information about long-term watershed behavior and control measure performance. The time needed to collect data describing the watershed and calibrate the model is low, probably on the order of a few person-days.

## 2.8 USEPA STORM WATER MANAGEMENT MODEL

The U.S. Environmental Protection Agency Storm Water Management Model, or SWMM, is a large, relatively complex software package capable of simulating the movement of precipitation and pollutants from the ground surface, through pipe/channel networks and storage/treatment facilities, and finally to receiving waters. The model can be used to simulate a single event or a long, continuous period.

SWMM has been released under several different "official" versions (Metcalf and Eddy, Inc. et al., 1971; Huber et al., 1975, 1984; Huber and Dickinson, 1988; Roesner et al., 1984, 1988) and there are many "unofficial" versions modified for specific purposes (some offered by private vendors). The original versions were designed for mainframe use, but the later versions can be executed on a personal computer. The current version of SWMM (Version 4; Huber and Dickinson, 1988 and Roesner et al., 1988) may be obtained (along with the documentation) from the Center for Exposure Assessment Modeling, U.S. Environmental Protection Agency, College Station Road, Athens, Georgia, 30613, 706-546-3549.

SWMM is probably the most popular of all urban runoff models. Unfortunately, it has a reputation for being a difficult model to use. This is not necessarily the case if one knows the fundamentals of how it works *and* if the parts of the model not needed in a particular application are discarded. SWMM uses well-known hydrologic and hydraulic concepts to simulate the urban watershed. Its reputation for sophistication (and difficulty) derives more from the numerical algorithms necessary to solve the rather straightforward governing equations that are trying to simulate a complex system (i.e., the urban watershed) driven by a highly dynamic input (i.e., precipitation).

There is an extensive body of literature describing SWMM's use and interested readers should begin their review of this literature by referring to a document prepared by Huber et al. (1985). This large body of experience is an advantage that SWMM probably enjoys over all other urban runoff models.

SWMM is divided into several "blocks". The major blocks, i.e., RUNOFF, TRANSPORT, STORAGE/TREATMENT, and EXTRAN, are *computational blocks* responsible for the hydrologic, pollutant generation and transport, and hydraulic calculations. Others, i.e., EXECUTIVE, STATISTICAL, RAIN, TEMP, GRAPH, AND COMBINE, perform various auxiliary functions, and are known as *service blocks*. A general operational schematic of SWMM is shown in Figure 22. Beginning with the computational blocks, each is described below.

SWMM is written in FORTRAN-77. While not very user friendly, it is not overly difficult to manage and use. Some "preprocessing" packages are is available to help prepare the input data, but most users will very quickly tire of these and simply prepare the data set with a word processor or other editing software.

### 2.8.1 RUNOFF Block

The RUNOFF Block generates surface runoff and pollutant loads in response to precipitation and surface pollutant accumulations (Huber and Dickinson, 1988). In many ways, RUNOFF is the "heart" of SWMM.

The key to applying RUNOFF is the division of the watershed into a number of subwatersheds (or subcatchments). Each subwatershed should be relatively

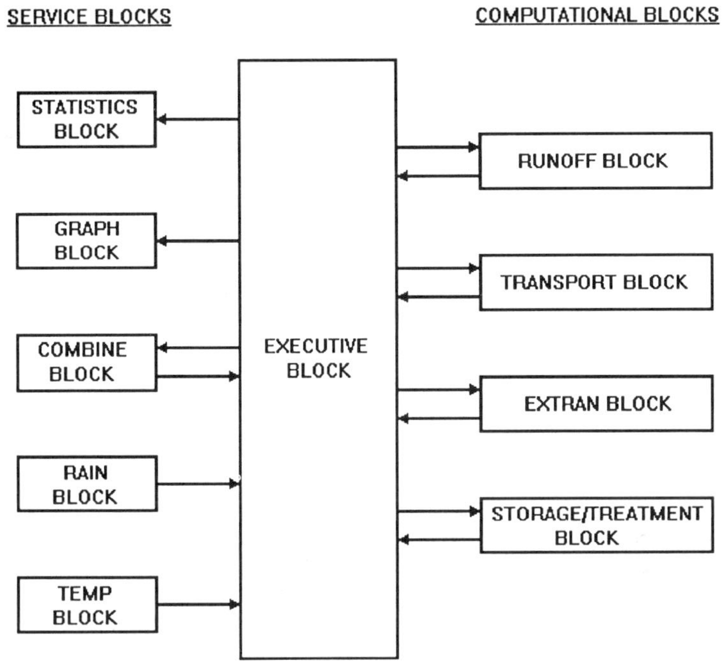


**Figure 22. Storm Water Management Model, program configuration (after Huber and Dickinson, 1988).**

homogeneous (i.e., the physical characteristics should be consistent). Just how homogeneous each subwatershed should be depends on how finely characterized the watershed must be to meet the modeling objectives. Dividing the watershed into a large number of subwatersheds implies that each is probably very homogeneous; a smaller number implies less homogeneity. This topic is discussed in more detail in Chapter 3.

*Runoff Simulation.* The conceptual view of surface runoff used by the RUNOFF Block is quite simple and is summarized in Figure 23. Essentially, each subwatershed surface is treated as a nonlinear reservoir with a single inflow, precipitation. There are several "discharges" including infiltration, evaporation, and surface runoff. The capacity of this "reservoir" is the maximum depression storage, which is the maximum surface storage provided by ponding, surface wetting, and interception. Surface runoff occurs only when the depth of water in the "reservoir" exceeds the maximum depression storage.

The water in storage is also being depleted by infiltration and evaporation. Infiltration occurs only if the ground surface is pervious (as opposed to an impervious surface, such as a paved parking lot, which by definition allows no infiltration). The infiltration process can be modeled by either of two methods (Horton's equation or the Green-Ampt equation). Infiltrated water is routed through upper and lower subsurface zones and may contribute to total runoff

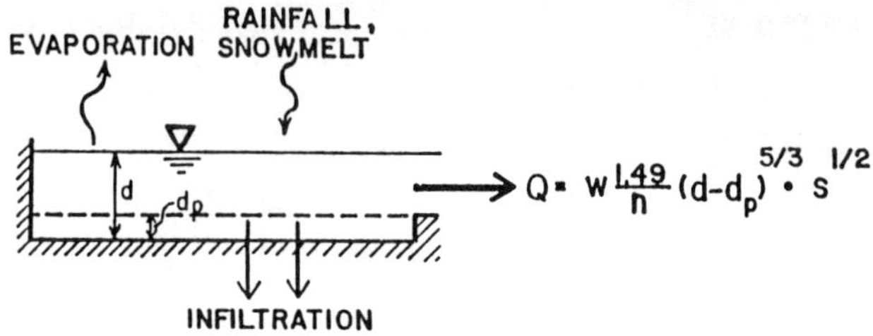


**Figure 23. Nonlinear reservoir representation of a subwatershed, RUNOFF Block, SWMM (from Huber and Dickinson, 1988).**

through ground water flow (this capability is a relatively new addition to SWMM). Monthly average evaporation rates (provided by the user) are used directly to evaporate water from the surface (and indirectly to calculate evapotranspiration from the subsurface zones). The precipitation intensity less the rates of infiltration and evaporation is the net inflow to the "reservoir."

The entire process is repeated for each subwatershed (each having its own unique set of physical characteristics) and is modeled by two equations. One is the continuity equation, which keeps track of the volume or depth of water on the surface of the subwatershed:

| change in volume stored on the subwatershed | = | rainfall excess (net inflow to the subwatershed) | – | runoff (outflow from the subwatershed) |
|---|---|---|---|---|
| $\frac{dV}{dt} = \frac{d(A \cdot d)}{dt}$ | = | $(A \cdot i_e)$ | – | $Q$ |

(2.27)

where
- $V$ = $A \cdot d$ = volume of water on the subwatershed, feet$^3$ or meters$^3$
- $A$ = area of the subwatershed, feet$^2$ or meters$^2$
- $d$ = depth of water on the subwatershed, feet or meters
- $t$ = time, seconds
- $i_e$ = rainfall excess, which is the rainfall intensity less the evaporation per infiltration rate, feet per second or meters per second
- Q = runoff flow rate from the subwatershed, feet$^3$ per second or meters$^3$ per second

The second equation is based on Manning's equation and is used to model the rate of surface runoff (i.e., the outflow rate from the reservoir) as a function of the depth of flow above the maximum depression storage depth. Manning's equation can be stated as

$$Q = A_c \cdot (\beta/n) \cdot R^{2/3} \cdot S_o^{\ 1/2} \tag{2.28}$$

where
- $A_c$ = cross-sectional area of flow over the subwatershed, feet$^2$ or meters$^2$
- $n$ = Manning's roughness coefficient
- $R$ = hydraulic radius of flow over the subwatershed, feet or meters
- $S_o$ = slope of the subwatershed, feet per foot or meters per meter (which is assumed to equal the friction or energy slope)
- $\beta$ = 1.49 if U.S. customary units are used or 1.0 if metric units are used

The cross-sectional area of flow is

$$A_c = W \cdot \left(d - d_p\right) \tag{2.29}$$

where
- $W$ = width of flow over the subwatershed, or the width of overland flow, feet or meters
- $d_p$ = depth of maximum depression storage, feet or meters

The hydraulic radius is the cross-sectional area of flow divided by the wetted perimeter. Since the depth of flow is very small, the wetted perimeter can be approximated by $W$. Thus, $R$ can be calculated as

$$R = \left[W \cdot \left(d - d_p\right)\right] / W = d - d_p \tag{2.30}$$

Substituting Equations 2.29 and 2.30 into Equation 2.28 yields

$$Q = W(\beta/n)\left(d - d_p\right)^{5/3} S_o^{1/2} \tag{2.31}$$

Substituting Equation 2.31 into Equation 2.27 and dividing by $A$ produces

$$\frac{dd}{dt} = i_e - \left[(\beta \cdot W)/(A \cdot n)\right] \cdot \left(d - d_p\right)^{5/3} \cdot S_o^{1/2} \tag{2.32}$$

Equation 2.32 is the second governing equation used in RUNOFF.

The two governing equations are solved numerically as follows. Equation 2.27 can be approximated by

$$\frac{d_{n+1} - d_n}{\Delta t} = \bar{i}_e - \bar{Q}/A \tag{2.33}$$

where $\Delta t$ = $t_{n+1} - t_n$, time step size, seconds
$n, n+1$ = subscripts indicating conditions at the end of time step n (or start of time step $n+1$) and the end of time step $n+1$ (e.g., $d_{n+1}$ is the depth at the end of time step $n+1$)
$\bar{i}_e$ = average precipitation intensity during time step $n+1$, feet per second or meters per second
$\bar{Q}$ = average runoff flow rate during time step $n+1$, feet$^3$ per second or meters$^3$ per second

Equation 2.33 shows the differential term $dd/dt$ is approximated by a finite difference of values for depth at two points in time separated by $\Delta t$. The value of the differential term is then approximated by the average of the terms on the right-hand side evaluated at the beginning and end of $\Delta t$. (See the discussion on numerical methods in Section 2.9.) If the average runoff flow rate is calculated as a function of the average depth of flow, Equation 2.33 becomes:

$$\frac{d_{n+1} - d_n}{\Delta t} = \bar{i}_e - [(\beta \cdot W)/(A \cdot n)] \cdot \left(\bar{d} - d_p\right)^{5/3} \cdot S_o^{1/2} \tag{2.34}$$

where $\bar{d}$ = $(d_n + d_{n+1})/2$, average depth of flow during time step $n+1$, feet or meters

Equation 2.34 is a relatively simple nonlinear, algebraic relationship with one unknown at any time, $d_{n+1}$. (The value of $d_n$ is, of course, known from the end of the previous time step.) The Newton-Raphson technique for numerically solving a nonlinear equation is used to solve for $d_{n+1}$. The calculated value of $d_{n+1}$ is then used in Equation 2.31 to calculate the value of $Q$ at the end of time step $n+1$. For all intents and purposes, Equation 2.34 is the core of the RUNOFF Block.

The most perplexing parameter in Equation 2.34 is the width of overland flow, $W$. Essentially, it is the width over which surface runoff occurs. Again, using the reservoir analogy, this width is similar to the length of a weir or spillway. An idealized view is shown in Figure 24. In this schematic, surface runoff is being discharged to a drainage channel running down the center of the subwatershed. In this situation the two halves are symmetrical and, thus, the total length of overland flow is twice the length of the channel. Of course, this idealized case never occurs, but it demonstrates the concept.

The width of overland flow primarily affects the rapidity of runoff. When the width of overland flow, or "weir," is enlarged, the length of the "reservoir" is shortened so that the surface area and depth of flow behind the "weir" remain constant for a given volume of water. As a result, a shorter width will delay

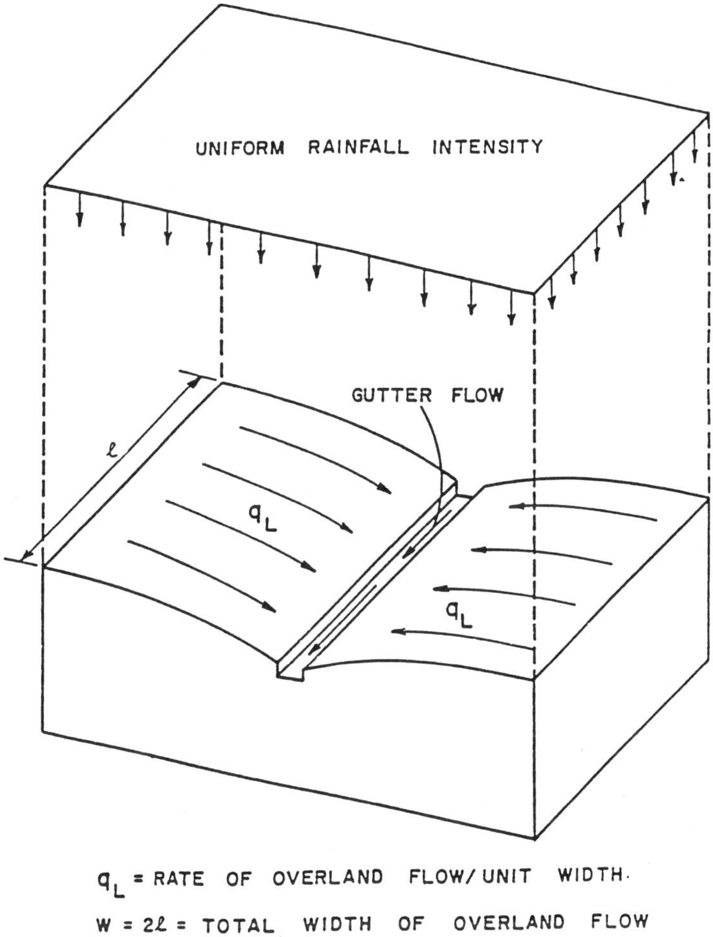


**Figure 24. Idealized subwatershed-gutter arrangement illustrating the subwatershed width of overland flow, RUNOFF Block, SWMM (from Huber and Dickinson, 1988).**

runoff; a longer width will facilitate runoff. This effect is illustrated in Figure 25.

The width of overland flow, while conceptually plausible, is difficult to estimate on a real watershed. Thus, it is typically used as a "calibration" parameter (others may prefer to apply a less euphemistic term). This is the fate often suffered by parameters that are more abstract.

The RUNOFF Block has a limited ability to route flows through simple gutters and pipes using the nonlinear reservoir technique. This feature will be used as part of an example in Chapter 3. However, the more sophisticated

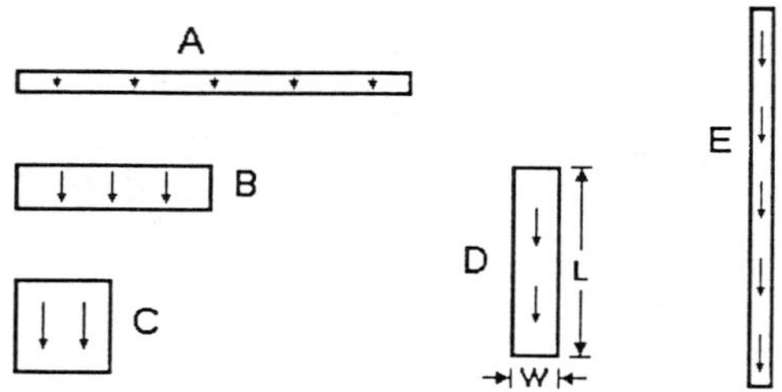


Rainfall: 1.0 in. / hr or 0.000023148 ft / s

Watershed Characteristics:
Area = 40,000 $ft^2$
Slope = 0.01 ft / ft or 1 %
Depression storage capacity = 0.0 in.
Imperviousness = 100 %
Manning's n for surface resistance = 0.02
Equilibrium runoff flow = rainfall intensity × area
= 0.926 cfs

| Shape | Width, W[a] (ft) | Length, L (ft) | Time of Concentration (min) |
|---|---|---|---|
| A | 800 | 50 | 3.7 |
| B | 400 | 100 | 5.7 |
| C | 200 | 200 | 8.6 |
| D | 100 | 400 | 13.0 |
| E | 50 | 800 | 19.7 |

a. Width of overland flow

Time Step: 5 min or 300 s

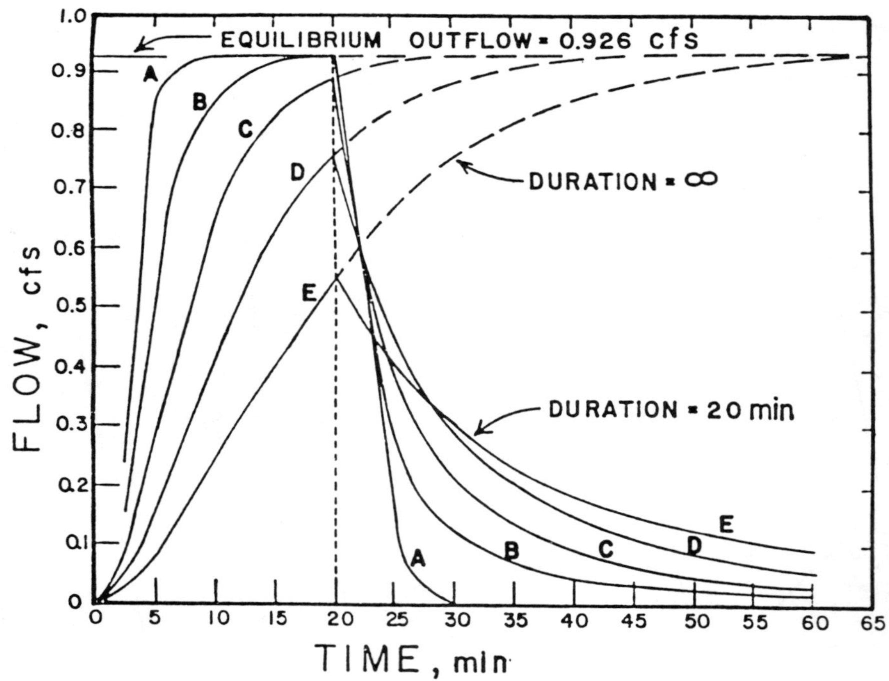


**Figure 25. Effect of the width of overland flow on runoff hydrographs, RUNOFF Block, SWMM (after Huber and Dickinson, 1988).**

routines in TRANSPORT and EXTRAN Blocks are almost always employed for this purpose.

The surface flows generated by the RUNOFF Block are concentrated at nodes. In other words, the flows are not distributed along gutters or pipes (as implied by Figure 24). The width of overland flow is used as a computational tool but the flow is not actually distributed over this distance.

*Pollutant Load Simulation.* The accumulation of pollutants on the subwatershed surface is modeled in a number of ways. Pollutants can be accumulated with "dust and dirt" on streets or as simple areal loads, much like the methods in STORM. However, whereas STORM accumulates surface loads at constant, or linear, rates, SWMM allows the rates of accumulation to be linear or nonlinear. The different methods (essentially four different equations) are summarized in Figure 26.

The washoff of accumulated pollutants can be modeled in two ways. One method is essentially the same as that used in STORM but it is formulated differently and it will be presented again in the context of SWMM. This method applies the following "first-order" relationship to each subwatershed:

$$-P_{off} = \frac{dP_p}{dt} = -K \cdot P_p \tag{2.35}$$

where $P_{off}$ = rate at which pollutant is washed off the subwatershed at time $t$, quantity per second
$P_p$ = amount of pollutant $p$ on the subwatershed surface at time $t$, quantity
$K$ = coefficient, 1 per second$^{-1}$

The term "quantity" is used in the definitions of $P_{off}$ and $P_p$ because the pollutants modeled by SWMM can be characterized with a variety of units (e.g., milligrams, MPN for coliform bacteria, JTU for turbidity, etc.). Equation 2.35 says that the rate at which a pollutant disappears from a subwatershed surface is proportional to the amount remaining on the subwatershed surface. As with STORM, the coefficient $K$ is assumed to be proportional to the runoff rate:

$$K = R_c \cdot r \tag{2.36}$$

where $R_c$ = washoff coefficient, inches$^{-1}$ or millimeters$^{-1}$
$r$ = runoff rate over the subwatershed at time $t$, inches per second or millimeters per second (calculated from $Q$ in Equation 2.31, $r = Q/A$)

Substituting Equation 2.36 into Equation 2.35 and multiplying by −1 yields

$$P_{off} = \frac{-dP_p}{dt} = R_c \cdot r \cdot P_p \tag{2.37}$$

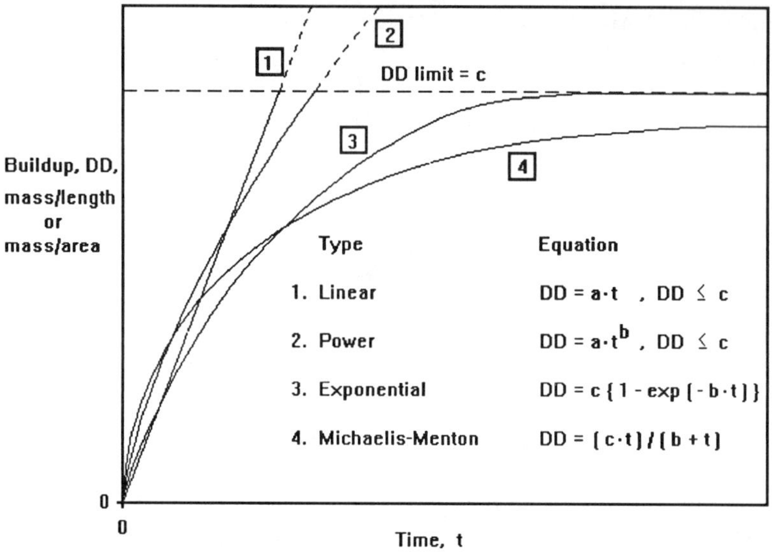


**Figure 26. Buildup equations, RUNOFF Block, SWMM (after Huber and Dickinson, 1988).**

A major deficiency of Equation 2.37 is that the runoff pollutant concentrations are forced to decrease over the course of a runoff event. Equation 2.37 shows the washoff rate increasing with runoff, but dividing Equation 2.37 by the runoff flow, $Q$, yields

$$C = \frac{P_{off}}{Q} = conv\frac{\left(-R_c \cdot r \cdot P_p\right)}{A \cdot r} = conv\frac{\left(-R_c \cdot P_p\right)}{A} \tag{2.38}$$

where $C$ = concentration, quantity per volume;
$Q$ = $A \cdot r$, runoff flow rate, feet$^3$ per second or meters$^3$ per second;
$A$ = subwatershed area, acres or hectares
$conv$ = a constant containing a number of conversion factors

Note that the runoff rate, $r$, disappears in Equation 2.38. Thus, the concentration, $C$, becomes independent of the runoff rate and directly proportional to a decreasing amount of pollutant remaining on the watershed. A decreasing concentration, while fairly common, is not the only possible trend. Concentrations can increase during a runoff event. To overcome this problem, an exponent other than 1 is allowed for $r$:

$$P_{off} = \frac{-dP_p}{dt} = R_c \cdot r^n \cdot P_p \tag{2.39}$$

where $n$ = exponent for the runoff rate

The load calculated by Equation 2.39 is combined with the runoff low rate to calculate the concentration, i.e., $C = P_{off}/Q$. If $n = 1$, Equation 2.39 reverts to Equation 2.37 and the concentration will decrease over the course of an event. Otherwise, concentration is proportional to $r^{n-1}$ (recall Equation 2.38) and as such it may increase if the runoff rate is large enough to offset the reduced value of $P_p$.

The solution to Equation 2.39 is determined from a finite difference approximation which produces

$$P_p(t+\Delta t) = P_p(t) \cdot \exp\left\{-R_c \cdot 0.5 \cdot \left[r(t)^n + r(t+\Delta t)^n\right] \cdot \Delta t\right\} \quad \textbf{(2.40)}$$

where $0.5[r(t)^n + r(t + \Delta t)^n]$ = average runoff rate over $\Delta t$, inches per second or millimeters per second

The second washoff method allows the user to simulate the washoff as a simple function of the runoff rate:

$$P_{off} = R_c \cdot Q^n \quad \textbf{(2.41)}$$

where coefficients $R_c$ and n are assigned particular values for each pollutant. In this method, the simulation of pollutant washoff may be totally independent of the amount accumulated on the surface (i.e., the washoff or delivered load is a function of the runoff flow rate only) or may be linked to the accumulated amount by not allowing the total load discharged during a particular storm to exceed the amount present on the surface at the beginning of the storm.

*Other Capabilities and Summary.* There are many other capabilities not discussed here, most notably snowmelt simulation. The RUNOFF Block consumes a considerable portion of the SWMM user's manual, making it seem more profound and difficult than it is. Recall that the bases of the RUNOFF Block are a very simple nonlinear reservoir representation of the surface runoff process and a first-order washoff process.

### 2.8.2 TRANSPORT Block

The TRANSPORT Block routes flows and pollutant loads through a sewer system (Metcalf and Eddy, Inc. et al., 1971; Huber and Dickinson, 1988). The flows and loads are generated by the RUNOFF Block (or some other program) and input to points throughout the system. TRANSPORT also has the ability to generate and route dry-weather or sanitary sewage flows. Hydrographs and pollutographs can also be manually introduced at various points in the system. The term "sewer system" refers to any form of pipe and/or channel system.

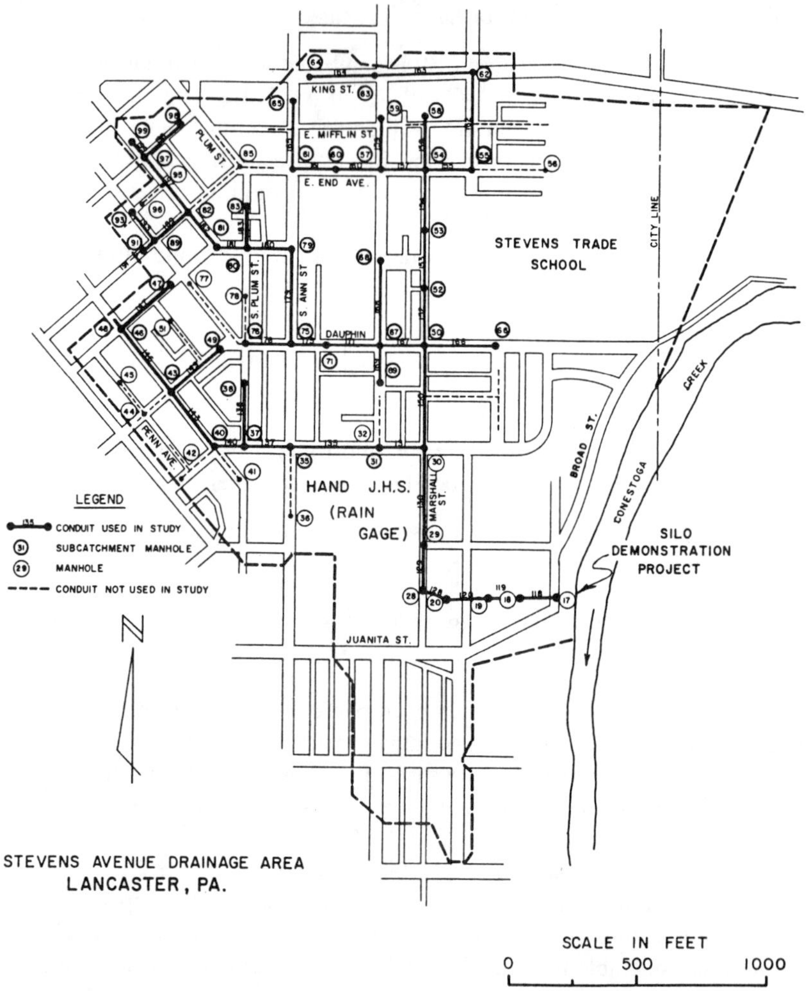


**Figure 27. Node and conduit representation of a sewer system, TRANSPORT Block, SWMM (Heaney at al., 1975).**

The sewer system is viewed as a series of "elements". This is shown in Figure 27. Elements may be nodes or conduits. Nodes link conduits and include manholes, pump stations, storage units, and flow dividers (see Table 4). Inflows to the system, such as surface runoff, occur at the nodes and may be entered directly by the user or come from other programs such as the RUNOFF Block. Each conduit may have one of 15 different cross-sectional shapes supplied by the model or 2 supplied by the user (see Table 4). Simple flow diversion devices (e.g., overflow structures) are also allowed.

Each element is identified by a user-supplied number. The numbering scheme can be arbitrary, for the system elements are fashioned into a connected

**Table 4. Elements in EXTRAN Block, SWMM (from Roesner et al., 1988)**

| Element Class | Types |
|---|---|
| Conduits/channels (links) | Circular |
| | Rectangular |
| | Phillips standard egg shape |
| | Boston horseshoe |
| | Gothic |
| | Catenary |
| | Louisville semielliptic |
| | Basket-handle |
| | Semicircular |
| | Modified basket-handle |
| | Rectangular, triangular bottom |
| | Rectangular, round bottom |
| | Trapezoid |
| | Parabolic |
| | Power Function |
| | HEC-2 Format — Natural Channel |
| | User supplied |
| Nonconduits (nodes) | |
| Manhole | Lift station |
| | Flow divider |
| | Storage unit |
| | Flow divider — weir |
| | Flow divider |
| | Backwater element |

network by indicating which elements are upstream of each element. In other words, element 11 is not necessarily connected to element 12 nor is element 12 necessarily connected to element 13. However, element 119 *can* be connected to element 1034 if the user specifies that element 1034 is one of the elements immediately upstream of element 119.

*Flow Routing*. Ideally, flow in sewers can be represented by two partial differential equations: the continuity and momentum equations or, as they are sometimes known, the Saint-Venant equations (Chow et al., 1988):

*Momentum*:

| pressure force | convective acceleration | local acceleration | gravity force | friction force |
|---|---|---|---|---|

$$\frac{\delta h}{\delta x} + \frac{v}{g}\cdot\frac{\delta v}{\delta x} + \frac{1}{g}\cdot\frac{\delta v}{\delta t} = S_o - S_f \qquad \textbf{(2.42)}$$

| *Continuity*: inflows and outflows to a control volume | | change in amount of water in control volume | | |
|---|---|---|---|---|
| $\frac{\delta Q}{\delta x}$ | $+$ | $\frac{\delta A}{\delta t}$ | $=$ | $0$ |

**(2.43)**

where
$h$ = water depth, feet or meters
$v$ = average flow velocity, feet per second or meters per second
$x$ = distance along the conduit, feet or meters
$t$ = time, seconds
$g$ = acceleration due to gravity, 32.2 feet per second$^2$ or 9.8 meters per second$^2$
$S_o$ = invert slope (slope of the conduit), feet per foot or meters per meter
$S_f$ = friction (energy) slope, feet per foot or meters per meter
$Q$ = flow rate, feet$^3$ per second or meters$^3$ per second
$A$ = cross-sectional area of flow, feet$^2$ or meters$^2$

Unfortunately, the Saint-Venant equations are difficult to manipulate and simplifications are often desirable. TRANSPORT uses a simplified version of the momentum equation in which all terms on the left hand side are neglected, i.e.,

$$S_f = S_o \tag{2.44}$$

The friction slope, $S_f$, is estimated from Manning's equation:

$$S_f = \frac{Q^2}{(\beta/n)^2 \cdot A^2 \cdot R^{4/3}} \tag{2.45}$$

where
$n$ = Manning's roughness coefficient
$R$ = hydraulic radius, feet or meters
$\beta$ = 1.49 when U.S. customary units are used or 1.0 when metric units are used

From Equations 2.44 and 2.45,

$$Q = (k/n) \cdot A \cdot R^{2/3} \cdot S_o^{1/2} \tag{2.46}$$

Essentially the original momentum equation, Equation 2.42, is replaced with an equation where $Q$ is a function of depth only (recall that the cross-sectional area, $A$, and the hydraulic radius, $R$, are functions of depth, $h$). This *kinematic wave approximation* and is what distinguishes the TRANSPORT Block from its more sophisticated cousin, the EXTRAN Block (see Section 2.8.3). Since flow is a function of depth alone, "disturbances" or changes that occur at one point in a sewer system can only affect what happens at downstream points, not upstream points. The full momentum equation propagates the effects of disturbances in both directions since flow (or velocity) is also characterized as a function of local and convective acceleration and pressure. With hydraulic effects propagated in only the downstream direction, backwater conditions cannot be simulated. In addition, the fact that flow is treated as a function of only depth means that the TRANSPORT Block cannot simulate surcharge conditions (flow under pressure). In summary, the TRANSPORT Block views the system as a simple cascade of conduits with downstream conduits having no effect on upstream conduits.

It is especially important to understand how the TRANSPORT Block behaves when it encounters surcharge conditions. Flows exceeding the open-channel capacity of a conduit are stored at the upstream end of the conduit (at a node) and released when the flow is less than this capacity. The flow in this situation is limited or "clipped" (as shown in Figure 28) and, as a result, downstream conditions are distorted.

The continuity equation, Equation 2.43, is approximated by a finite difference relationship:

$$\frac{(1-w_t)(A_{j,n+1}-A_{j,n})+w_t(A_{j+1,n+1}-A_{j+1,n})}{\Delta t}+\frac{(1-w_x)(Q_{j+1,n}-Q_{j,n})+w_x(Q_{j+1,n+1}-Q_{j,n+1})}{\Delta x}=0 \tag{2.47}$$

where
- $\Delta t = t_{n+1} - t_n$, time step size, seconds
- $\Delta x = x_{j+1} - x_j$, distance interval length (the conduit length), feet or meters
- $j, j+1$ = subscripts indicating conditions at the upstream end and the downstream end of conduit $M$, respectively
- $n, n+1$ = subscripts indicating conditions at the end of time step $n$ (which is also the beginning of time step $n+1$) and the end of time step $n+1$, respectively
- $w_t, w_x$ = weights

The weights $w_x$ and $w_t$ are both set to 0.55 . (This is the result of a series of tests to determine the best values for numerical stability. See Metcalf and Eddy, Inc., et al., 1971.) This numerical approximation is illustrated in Figure 29.

Equations 2.46 and 2.47 work together to route flows through a sewer system. At the end of any time step $n+1$, the unknowns quantities are the flow and cross-sectional area of flow at the downstream end of conduit $M$, $Q_{j+1,n+1}$, and $A_{j+1,n+1}$. (The variables $Q_{j,n}$ and $A_{j,n}$ are known from the previous time step

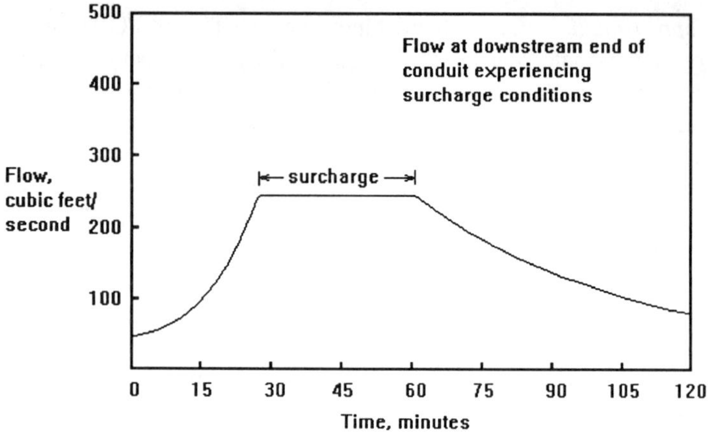


**Figure 28. Effect of surcharging on downstream hydrograph, TRANSPORT Block, SWMM.**

and conditions at the upstream end of the conduit.) With only two unknowns, these two equations are sufficient to determine the value of both. The calculations are carried out from the most upstream conduit to the most downstream conduit during each time step.

Nodes (e.g., manholes) are treated very simply in the TRANSPORT Block. Flow exiting a node is just the sum of all the flows entering the node.

Although the numbering scheme used to identify the various elements is arbitrary from the user's perspective, the program establishes a separate internal numbering scheme for the routing calculations. Again, this is possible because the user identifies the elements upstream of each element.

It should be noted that several modifications are made to Equations 2.46 and 2.47 to improve model accuracy. These will not be discussed here. It is sufficient to say that these two equations are the heart of the TRANSPORT Block.

*Pollutant Routing*. Pollutants are routed through the system by treating each conduit as a completely mixed reactor with first-order decay. The governing differential equation is shown below:

| change in mass in conduit per | mass rate to conduit | mass rate from conduit | decay in conduit | mass source or sink |
|---|---|---|---|---|

$$\frac{dVC}{dt} = \frac{V \cdot dC}{dt} + \frac{C \cdot dV}{dt} = (Q_i \cdot C_i) - (Q \cdot C) - K \cdot C \cdot V \pm L \tag{2.48}$$

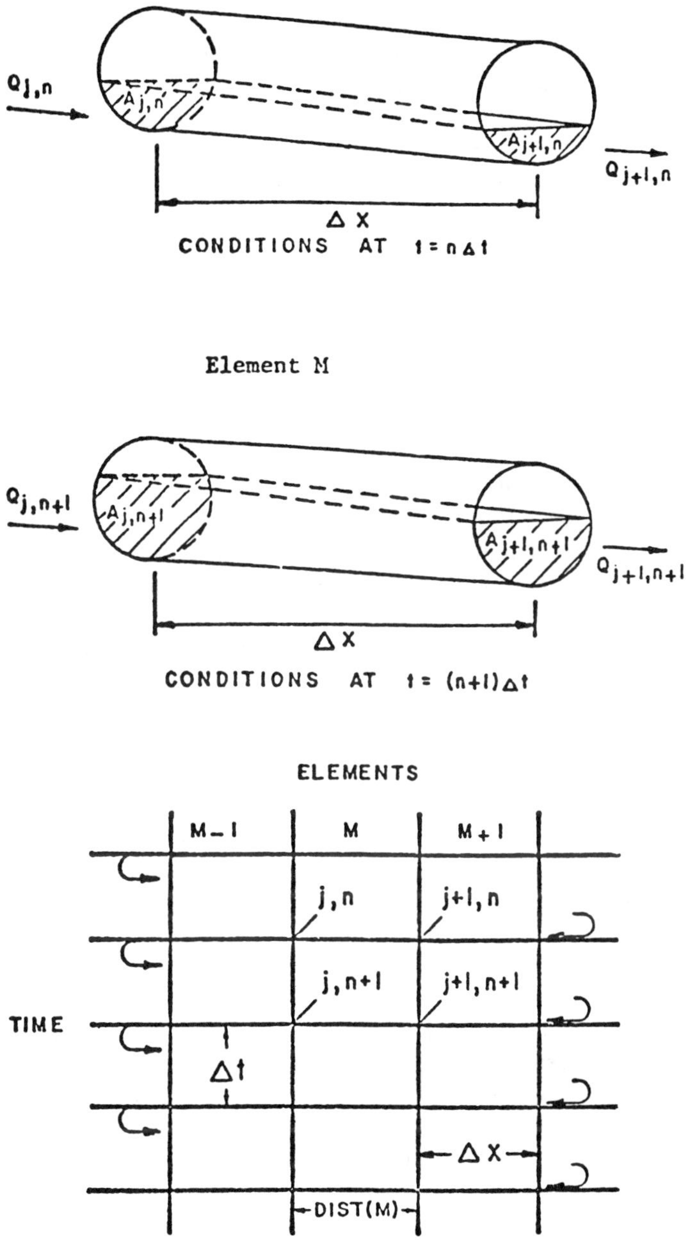


**Figure 29. Numerical approximation definitions in TRANSPORT Block, SWMM (from Metcalf and Eddy, Inc. et al., 1971).**

where $C$ = pollutant concentration in conduit and discharge from conduit, quantity per unit volume
$V$ = volume of water in conduit, feet$^3$ or meters$^3$
$Q_i$ = inflow rate to conduit, feet$^3$ per second or meters$^3$ per second
$C_i$ = pollutant concentration in inflow, quantity per unit volume
$Q$ = outflow rate from conduit, cfs or meters$^3$ per second
$K$ = first-order decay coefficient, seconds$^{-1}$
$L$ = source (or sink) of pollutant to the conduit, quantity per second

An integrated form of the solution to this differential equation is used with a simple numerical technique to carry out the estimates for concentration in each conduit.

*Other Capabilities and Summary.* The TRANSPORT Block has very simple routines to estimate infiltration into the sewer system. The routine is not very useful and the user can just as easily enter infiltration flows at various nodes in the system. Unfortunately, there is no direct way to generate sewer infiltration from the subsurface flows simulated in the RUNOFF Block.

The TRANSPORT Block also contains a rather data-intensive routine for estimating dry-weather flows and pollutant loads (which would be useful in watersheds with combined sewer systems). The estimates are calculated as functions of land use, population, income levels, and a host of other factors.

In summary, the TRANSPORT Block effectively routes flows and pollutants through a simple sewer system, as long as surcharging is not encountered. Unlike its companion, the EXTRAN Block, TRANSPORT is capable of routing pollutants. In most applications, it is numerically stable and easy to apply.

### 2.8.3 EXTRAN Block

The EXTRAN Block extends the hydraulic capabilities of the TRANSPORT Block (Roesner et al., 1988). Pollutant routing is not part of EXTRAN as of this writing, but may be added in the very near future. EXTRAN has a developmental history that is a little different from the rest of SWMM and only joined the software bundle in the later versions.

*Flow Routing.* As in the TRANSPORT Block, the sewer system is viewed as a network of links and nodes (or, collectively, elements). Inflows to the system occur at the nodes and may be entered directly by the user or come from the RUNOFF Block or other programs. The number of conduit types that can be modeled is not as extensive as that of the TRANSPORT Block and the method of linking the system together is slightly different (see Table 5). Because hydraulic "signals" are propagated in both directions, upstream *and* downstream nodes are identified for each link (or conduit).

**Table 5. Elements in TRANSPORT Block, SWMM (from Huber and Dickinson, 1988)**

| Element Class | Types |
|---|---|
| Conduits or links | Rectangular |
| | Circular |
| | Horseshoe |
| | Eggshape |
| | Baskethandle |
| | Trapezoid |
| | Power function |
| | Natural Channel (irregular cross section) |
| Junctions or nodes | Manholes |
| Diversion structures | Orifices |
| | Transverse weirs |
| | Side-flow weirs |
| Pump stations | On-line or off-line pump station |
| Storage basins | On-line, enlarged pipes or tunnels |
| | On-line or off-line, arbitrary stage-area relationship |
| Outfall structures | Transverse weir with tide gate |
| | Transverse weir without tide gate |
| | Side-flow weir with tide gate |
| | Side-flow weir without tide gate |
| | Outfall with tide gate |
| | Free outfall without tide gate |

The EXTRAN Block uses the complete Saint-Venant equations to model the routing of flows. However, the equations are expressed a little differently from those in the TRANSPORT Block (i.e., Equations 2.42 and 2.43):

*Momentum*: pressure and gravity force + convective acceleration + local acceleration + friction force

$$g \cdot A \cdot \frac{\delta H}{\delta x} + \frac{\delta\left(Q^2/A\right)}{\delta x} + \frac{\delta Q}{\delta t} + g \cdot A \cdot S_f = 0 \quad \textbf{(2.49)}$$

*Continuity*: inflows and outflows to a control volume + change in amount of water in control volume

$$\frac{\delta Q}{\delta x} + \frac{\delta A}{\delta t} = 0 \quad \textbf{(2.50)}$$

where $H$ = $z + h$, hydraulic head, feet or meters
$z$ = conduit invert elevation, feet or meters
$h$ = water depth, feet or meters
$x$ = distance along the conduit, feet or meters
$t$ = time, seconds
$g$ = acceleration due to gravity, 32.2 feet per second$^2$ or 9.8 meters per second$^2$
$S_f$ = friction (energy) slope, feet per foot or meters per meter
$Q$ = flow rate, feet$^3$ per second or meters$^3$ per second
$A$ = cross-sectional area of flow, feet$^2$ or meters$^2$

The friction force term is estimated by Manning's equation:

$$S_f = \frac{Q^2}{(\beta/n)^2 \cdot A^2 \cdot R^{4/3}} = \frac{Q \cdot |v|}{(\beta/n)^2 \cdot A \cdot R^{4/3}} \tag{2.51}$$

where $n$ = Manning's roughness coefficient
$R$ = hydraulic radius, feet or meters
$\beta$ = 1.49 when U.S. customary units are used or 1.0 when metric units are used

The absolute value sign on velocity, $v$, insures that the friction force, as expressed by $S_f$, opposes the direction of flow. For example, if flow is reversed (from the nominal direction of flow) both $Q$ and $v$ would be negative. Taking the absolute value of $v$ allows $S_f$ to become negative as well.

Equations 2.49 and 2-50 are combined through a few algebraic manipulations. The first of these relies on the identity

$$Q^2/A = v^2 A \tag{2.52}$$

where $v$ = average flow velocity, feet per second or meters per second

Substituting Equation 2.52 into the convective acceleration term of the momentum equation (Equation 2.49) yields

$$g \cdot A \cdot \frac{\delta H}{\delta x} + 2A \cdot v \cdot \frac{\delta v}{\delta x} + v^2 \cdot \frac{\delta A}{\delta x} + \frac{\delta Q}{\delta t} + g \cdot A \cdot S_f = 0 \tag{2.53}$$

Noting that $Q = A \cdot v$, the continuity equation, Equation 2.50, can be written as

$$A \cdot \frac{\delta v}{\delta x} + v \cdot \frac{\delta A}{\delta x} + \frac{\delta A}{\delta t} = 0 \tag{2.54}$$

Multiplying by velocity $v$ and rearranging terms yields

$$A \cdot v \cdot \frac{\delta v}{\delta x} = -v \cdot \frac{\delta A}{\delta t} - v^2 \cdot \frac{\delta A}{\delta x} \tag{2.55}$$

Substituting this result into the second term of the revised momentum equation (Equation 2.53) leads to the basic flow equation used in the EXTRAN Block:

$$g \cdot A \cdot \frac{\delta H}{\delta x} - 2v \cdot \frac{\delta A}{\delta t} - v^2 \cdot \frac{\delta A}{\delta x} + \frac{\delta Q}{\delta t} + g \cdot A \cdot S_f = 0 \tag{2.56}$$

Essentially, Equation 2.56 contains two variables, $Q$ and $H$ ($v$ and $A$ are related to $Q$ and $H$). Therefore, the continuity equation (Equation 2.50) is used to provide a second equation relating $Q$ and $H$. Finite difference approximations are used to numerically solve the two partial differential equations. The details will not be discussed here. The numerical techniques used in the EXTRAN Block are somewhat unstable and some attention must be paid to the size of the time step and conduit lengths (see Section 2.9 for a discussion on numerical methods and stability).

*Other Capabilities and Summary.* The routing of pollutant loads is not currently modeled. Nor are there routines for estimating sewer infiltration or dry-weather flows.

The EXTRAN Block should be used with care and not taken lightly. While hydraulically powerful, it has proven to be numerically "temperamental". Nevertheless, EXTRAN should be used if the sewer system to be modeled is complex and subject to surcharging.

### 2.8.4 STORAGE/TREATMENT Block

The STORAGE/TREATMENT Block is designed to route flow and pollutant loads through a storage/treatment facility (Nix, 1982; Huber and Dickinson, 1988). These flows and loads may come from other blocks in SWMM or other sources. The user is given a great deal of flexibility by the block's ability to connect as many as five storage/treatment units together in a variety of networks. Each unit may be given detention (or storage) characteristics or be modeled as a simple flow-through device.

If a unit is modeled as a detention unit, as shown in Figure 30, flows are routed through the unit with a level-surface flow routing procedure (i.e., the modified Puls method). This method is based on yet another version of the continuity of mass equation (Viessman et al., 1989):

| change in volume of water in detention unit per unit time | | flow rate entering the detention unit | | flow rate leaving the detention unit |
|---|---|---|---|---|
| $\frac{dV}{dt}$ | = | $I$ | − | $Q$ |

$$\frac{dV}{dt} = I - Q \tag{2.57}$$

where $V$ = volume of water in detention unit, feet$^3$ or meters$^3$
$I$ = inflow rate, feet$^3$ per second or meters$^3$ per second
$Q$ = outflow rate, feet$^3$ per second or meters$^3$ per second
$t$ = time, $s$

Equation 2.57 is approximated by the following finite difference relationship:

$$\frac{\left(V_{n+1} - V_n\right)}{\Delta t} = \frac{\left(I_n + I_{n+1}\right)}{2} + \frac{\left(Q_n + Q_{n+1}\right)}{2} \tag{2.58}$$

where $n$, $n+1$ = subscripts indicating conditions at the end of time step $n$ (or the beginning of time step $n+1$) and the end of time step $n+1$, respectively
$\Delta t$ = $t_{n+1} - t_n$, time step size, seconds

The values of $V_{n+1}$ and $Q_{n+1}$ are unknown at the end of time step $n+1$. (The values for $V_n$ and $Q_n$ are known from the previous time step.) Therefore, a second relationship between storage, $V$, and discharge, $Q$, is needed to determine their values. The program gives the user two ways to provide this relationship. One uses a linear interpolation algorithm to approximate the relationship through a series of volume-discharge data pairs (each pair occurring at a particular depth). The second method uses an exponential function to describe the relationship.

Pollutants are routed through the detention unit in either a completely mixed or plug-flow manner. In the completely mixed case all incoming material is instantly distributed throughout the detention unit and, thus, the pollutant concentration is uniform throughout the unit (see Figure 31). The following equation is used to simulate the fate of pollutants in the completely mixed detention unit:

| change in mass in detention basin per unit time | | mass rate entering the detention basin | | mass rate leaving the detention basin | | "reaction" of pollutant by first - order decay |
|---|---|---|---|---|---|---|
| $\frac{d(CV)}{dt}$ | = | $I \cdot C_I$ | − | $Q \cdot C$ | − | $K_c \cdot C \cdot V$ |

$$\frac{d(CV)}{dt} = I \cdot C_I - Q \cdot C - K_c \cdot C \cdot V \tag{2.59}$$

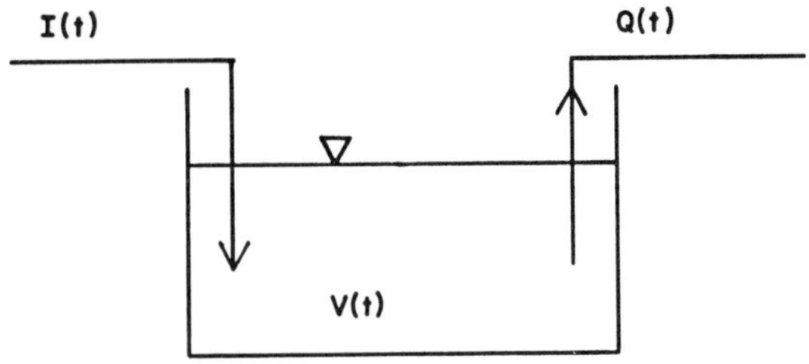


**Figure 30. Level-surface reservoir, STORAGE/TREATMENT Block, SWMM (after Huber and Dickinson, 1988).**

where $C_I$ = influent pollutant concentration, quantity per unit volume
$C$ = effluent pollutant concentration, quantity per unit volume
$K_c$ = decay coefficient, $\text{seconds}^{-1}$

Equation 2.59 is approximated by a finite difference equation in a manner very similar to that done for Equation 2.57. The result is an algebraic solution for the effluent pollutant concentration at the end of every time step.

In the plug-flow case, the stormwater and pollutants entering the detention unit in a given time step forms a "plug" (see Figure 32). The number of plugs (and/or fraction of a plug) leaving the unit in any time step is, of course, directly related to the departing volume (as determined by the flow routing procedure).

Pollutant removal is modeled through the use of "removal equations" or through a set of relationships describing discrete particle settling. In the former the program provides several variables (such as detention time, inflow rate, etc.) around which the user can build a wide range of removal equations. This is done by providing a generic function that can be manipulated through the assignment of the program variables to the variables in the generic equation and the selection of appropriate values for the equation coefficients. The generic function is

$$R = \begin{bmatrix} a_{12} \cdot \exp(a_1 x_1) \cdot x_2^{a_2} + a_{13} \cdot \exp(a_3 x_3) \cdot x_4^{a_4} + \\ a_{14} \cdot \exp(a_5 x_5) \cdot x_6^{a_6} + a_{15} \cdot \exp(a_7 x_7 + a_8 x_8) \cdot x_9^{a_9} \cdot x_{10}^{a_{10}} \cdot x_{11}^{a_{11}} \end{bmatrix}^{a_{16}} \qquad \textbf{(2.60)}$$

where $x_i$ = removal equation variable
$a_j$ = coefficients
$R$ = removal fraction, $0 \le R \le 1.0$

As mentioned above, each variable $x_i$ can represent one of a number of variables in the storage/treatment algorithm. The selection varies depending on whether the detention basin is assumed to behave as a plug-flow reactor or a completely mixed reactor. In the completely mixed mode Equation 2.60 is really only used to provide a value for $K_c$ in Equation 2.59. In the plug-flow mode the equation is applied to each plug and many more options are available.

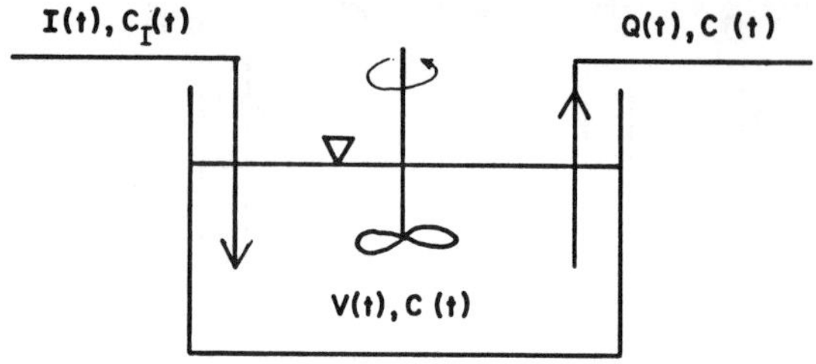


**Figure 31. Completely mixed detention unit, STORAGE/TREATMENT Block, SWMM (from Huber and Dickinson, 1988).**

The particle settling algorithm can only be used in the plug-flow mode. The size distribution of particles entering the detention unit is assumed to remain constant for all incoming flows. The settling of particles is based on the theory of discrete particle settling modified for the effects of turbulence (Nix, 1982). The outgoing size distribution changes over time as differences in flow conditions dictate. The removal (settling) of a particular pollutant is based on its association with particles of various settling velocities or sizes and specific gravities (e.g., 20% of the BOD load is associated with particles that have a given range of settling velocities). This association also remains constant for all incoming pollutant loads. This assumption and that of a constant particle size distribution are major limitations. It should be said, however, that these limitations exist mainly because the RUNOFF Block does provide information about particles in stormwater. The settling algorithms in the STORAGE/TREATMENT Block could handle time-varying distributions.

When a unit is defined as a simple flow-through or nondetention device, flow is routed without delay, i.e., inflow = outflow. Pollutant removal is simulated with Equation 2.60 (again, built by the user with variables provided by the program) or by assuming that all particles of a certain size or larger are removed.

The STORAGE/TREATMENT Block is not intended to be a sophisticated unit operations simulator. There are other models that simulate these processes in great detail. This block is designed to give the user a reasonable prediction of how a wet-weather facility will respond to dynamic stormwater flows and pollutant loads. In order to keep the model tractable, the representation of pollutant routing and removal is fairly simple.

### 2.8.5 Other Blocks

*Receiving Water Modeling*

Earlier versions of SWMM included an algorithm, known as the RECEIVE Block, capable of routing flows and pollutant loads through receiving waters such as rivers, streams, and estuaries (Metcalf and Eddy, Inc. et al., 1971;

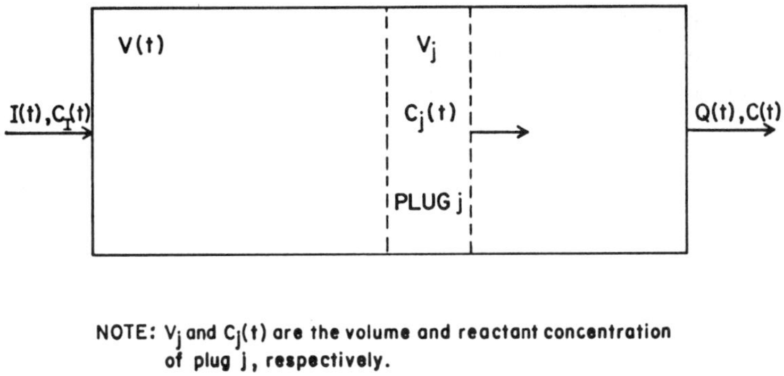


**Figure 32. Plug-flow detention unit, STORAGE/TREATMENT block, SWMM (from Huber and Dickinson, 1988).**

Huber et al., 1975). The latest version of SWMM does not contain the RECEIVE Block. However, there are provisions in two EPA receiving water models, DYNHYD4 and WASP4, for interfacing with SWMM (Ambrose et al., 1988).

### *EXECUTIVE Block*

The EXECUTIVE Block acts as the "traffic director" of SWMM. It directs the order in which the other blocks are executed and provides an interface between them, as well as external programs and data files. The interface files (containing the information that must be transferred between blocks) have a standardized format to ease the interaction between blocks and external programs. SWMM has been successfully interfaced with several other models.

### *GRAPH Block*

The GRAPH Block provides a limited ability to plot actual and simulated hydrographs, pollutographs, etc. SWMM suffers from the poor graphics capability of FORTRAN.

### *COMBINE Block*

The computational blocks of SWMM have limitations on the number of elements (pipes, subcatchments, etc.) they can simulate. The COMBINE Block allows the modeling of larger systems by "collating" and "combining" interface files. Collation refers to the block's ability to add flows and pollutant loads generated by two other blocks (or two runs of the same block) at common nodes (i.e., manholes, inlets, etc.). To combine means that the output from two blocks (at two different nodes) are added and placed at one node.

A limitation of the COMBINE Block is that some hydraulic "connectivity" may be lost when combining or collating interface files produced by the EXTRAN Block. Recall that the EXTRAN Block has the ability to route flows dynamically which essentially means that hydraulic effects can be propagated in the downstream and upstream directions. When two parts of a sewer system are modeled separately, the effects of a downstream run are lost to the upstream run.

*RAIN Block*

The RAIN Block processes lengthy precipitation records for use by the RUNOFF Block. These records are usually read from National Weather Service tapes and diskettes, although user-prepared records may also be processed. RAIN also incorporates an algorithm to perform storm event analyses. This routine is based on the well-known SYNOP program developed by Hydroscience, Inc. (1976, 1979).

*TEMP Block*

The TEMP Block processes daily temperature, evaporation, and wind speed data for use by the RUNOFF Block. These records are usually read from National Weather Service tapes and diskettes, although user-prepared records may also be processed.

*STATISTICS Block*

The STATISTICS Block performs a limited number of statistical operations on a sequence of events (rainfall or runoff events). A variety of quantity and quality parameters may be analyzed. Output can include (at the user's option) a table showing the sequential series of events, a table of listing the magnitude, return period, and frequency of the events, graphs showing magnitude vs. return period and frequency, and the first three moments of the event data.

### 2.8.6 Typical Application

The City of Rutland, VT comprises an area of approximately 7 square miles and a population of about 20,000. Initial construction of the sewer system began in the 1860s. The major features of the system as it is today are shown in Figure 33. This following summary comes from a paper by Nix et al. (1991).

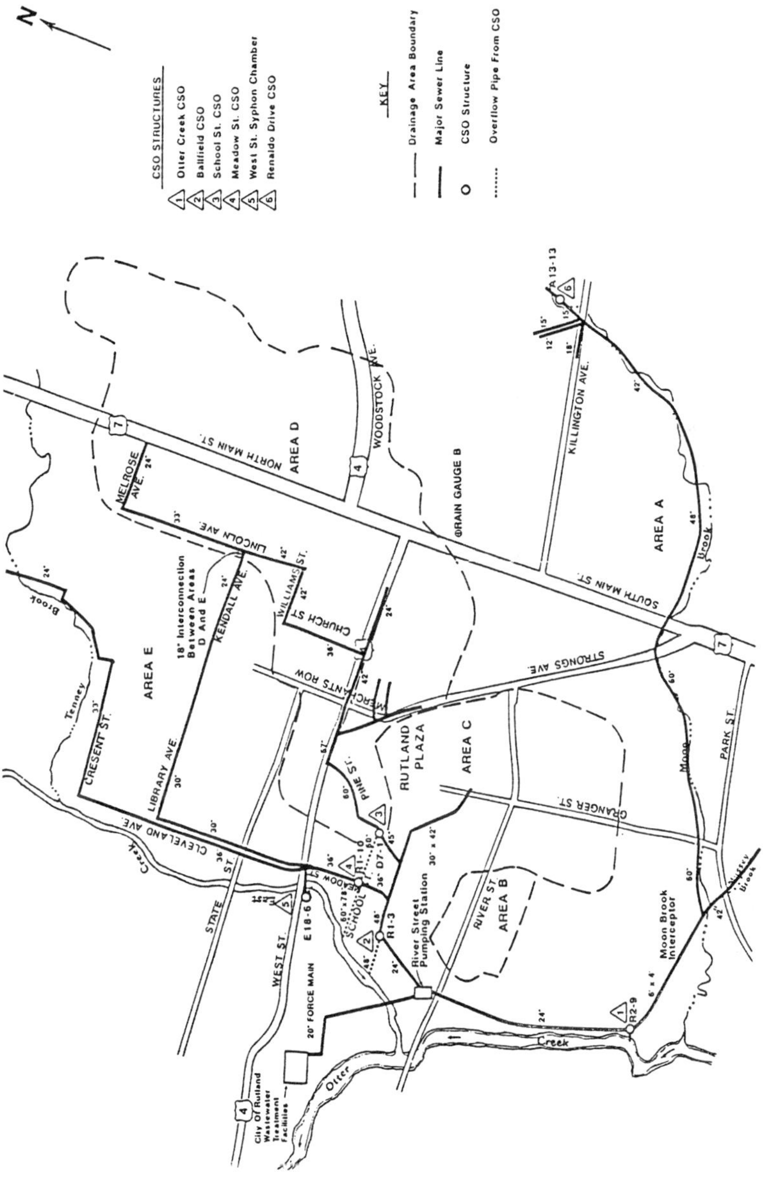


**Figure 33. Sewer system, Rutland, VT (from Nix et al., 1991).**

Prior to the construction of a sewage treatment plant in 1963, all combined sewage (sanitary sewage and urban runoff) was released directly to Otter Creek and its tributary, East Creek. Two 24-inch interceptors were also constructed at this time to transport the dry-weather sewage to a central pumping station (River Street Pumping Station) for transport through a force main to the treatment plant. However, wet-weather overflows from the system continue to plague the receiving waters.

The sewage system is divided into five drainage areas: areas A, B, C, D, and E. Areas B, C, D, and E comprise what is also known as the East Creek Drainage Area (as all wet-weather sewer overflows are discharged to East Creek). This area is dominated by older combined sewers. Area A is the largest drainage area, comprising roughly half of the city's acreage. The area is served by the Moon Brook sewer and its tributary, the Mussey Brook sewer. Considerable parts of this mostly residential area are served by separate sewers (i.e., separate storm and sanitary sewers) of fairly recent construction.

SWMM was used to establish a system model with the objective being to analyze the frequency of wet-weather overflows and to help determine abatement strategies. The areas contributing surface runoff to the sewer system were characterized by 22 subcatchments for use in the RUNOFF Block (8 were located in area A, 1 in area C, 6 in area D, and 7 in area E). Aerial photographs and 7.5-minute USGS topographic maps were used to provide the physical characteristics of the subcatchments. With a few exceptions, only conduits having diameters larger than 24 inches were used to characterize the sewer system for the EXTRAN Block. This level of discretization was chosen to ensure that all major features of the system were included while at the same time keeping the model size and complexity to a minimum. Sanitary sewage flows were included by adding constant flows at various points in the system. The completed model contained 110 elements (57 junctions and 53 conduits). A system of this size and detail is quite modest and well suited to the EXTRAN Block. Pollutants were not simulated (but this does *not* mean they were ignored).

The RUNOFF Block was executed with a time step of 5 minutes. This value was compatible with the "response time" of the watershed, and produced very small continuity errors and a smaller time step was considered unnecessary. A 15-second time step was used to execute the EXTRAN Block. This value represented a good compromise between stability considerations and run times. (See Section 2.9 for a discussion on numerical approximations and model stability.)

In order to validate the model simulated hydrographs were compared to measured hydrographs at a number of overflow sites for several storm events. Model parameters were adjusted so that there was as much agreement as possible with the resources available. Figure 34 shows the computed and actual results at one site (overflow #1) for a single storm. Once a validated model was achieved the model was used to predict the peak carrying capacity of the sewer system, a ranking of the relative frequency and quantity of overflows, locations of potential street flooding, and pollutant loadings based on flow projections and measured water quality data. This information has been used to evaluate various abatement alternatives.

### 2.8.7 Summary

The EPA Storm Water Management Model is among the most popular of urban stormwater software packages, if not *the* most popular. It was designed to build detailed models of urban watersheds and is capable of routing flows over the surface and through sewer and storage/treatment systems with considerable temporal and spatial detail. While it was originally intended to be a single-event simulator, it can be used effectively for continuous simulation. It has been used, for example, to develop the same types of long-term storage/treatment performance curves discussed in Sections 2.6.3 and 2.7.5 (Nix and Heaney, 1988; Nix et al., 1988).

While known as a large and complex model, SWMM simulates runoff and pollutant loads with fairly simple, flexible algorithms. In fact, SWMM is flexible enough to serve a number of different roles. It can be used to build a simple model of an urban watershed much as STORM does or it can create a detailed model designed to produce hydrographs and pollutographs at a number of locations in the watershed.

Users will find that, if they look carefully at what they are trying to accomplish, the application of SWMM can be rather straightforward. SWMM tries to be comprehensive and as a result it can appear to be somewhat cumbersome. However, most applications involve small portions of the package, so this appearance is deceiving. It also suffers from a rather old-fashioned user interface with some vestiges of the old "card input" days. Nevertheless, SWMM is a very good, time-tested piece of software.

An annotated input data guide for SWMM is presented as an appendix to this book.

## 2.9 NUMERICAL METHODS

Many urban runoff models, the more sophisticated ones in particular, use numerical methods to solve the differential equations describing the physical processes. This was illustrated for a few cases in Sections 2.7 and 2.8. Numerical methods, in general, convert complicated mathematical problems into simpler problems. In urban runoff models, we are mostly interested in converting differential equations into simple algebraic relationships that are readily solved by the digital computer.

The discussion here is not intended to present a complete tutorial on numerical methods or to bore those familiar with them. The goal is to make sure the

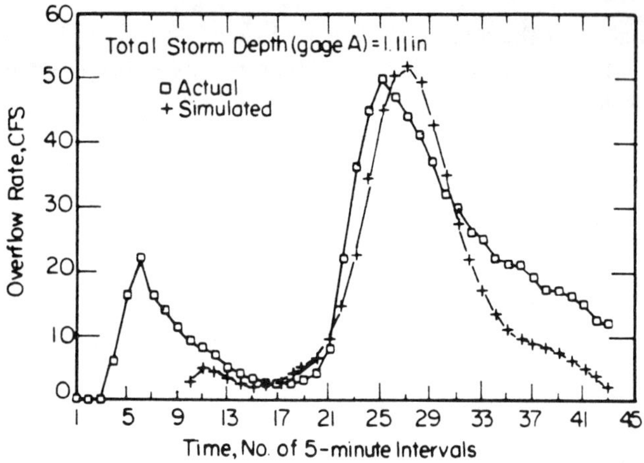


**Figure 34. Overflow hydrograph, overflow #1, Rutland, VT (Nix et al., 1991).**

reader appreciates the common mathematical manipulations and approximations and understands some of the problems that can arise as a result of their use.

One of the simplest of numerical methods approximates the ordinary differential equation

$$\frac{dy}{dx} = f(x, y) \tag{2.61}$$

with the finite difference form

$$\frac{y_{j+1} - y_j}{\Delta x} = \frac{\left[f(x_j, y_j) + f(x_{j+1}, y_{j+1})\right]}{2} \tag{2.62}$$

or

$$y_{j+1} = y_j + (\Delta x/2) \cdot \left[f(x_j, y_j) + f(x_{j+1}, y_{j+1})\right] \tag{2.63}$$

where $y$ = dependent variable to be solved
$x$ = independent variable
$\Delta x$ = $x_{j+1} - x_j$, finite interval of $x$
$j, j+1$ = subscripts indicating conditions at the low and high ends of interval $\Delta x$

Equation 2.62 is known as the second-order Adams formula. (A cursory look at the equation reveals that it is also an application of the trapezoidal rule for integration.) At any interval $\Delta x$, the known values are $x_j$, $x_{j+1}$, and $y_j$; the unknown value is $y_{j+1}$. The independent variable might, for example, be distance or time. This approximation is depicted in Figure 35. Essentially, the differential term (i.e., the slope) is estimated as an average of the derivatives at $x_j$ and $x_{j+1}$, which are simply the values of the function f at $x_j$ and $x_{j+1}$. Note that in Figure 35(a) the size of $\Delta x$ is small and as a result the approximation is relatively accurate. In Figure 35(b), the size of $\Delta x$ is large and the approximation suffers.

Another simple numerical method approximates Equation 2.61 by

$$\frac{y_{j+1} - y_j}{\Delta x} = f\left(x_j, y_j\right) \tag{2.64}$$

or

$$y_{j+1} = y_j + \Delta x \cdot f\left(x_j, y_j\right) \tag{2.65}$$

This technique, known as Euler's method, calculates a new estimate for $y$, $y_{j+1}$, based on projections of the derivative from the previous value of $y$, $y_j$.

Under certain conditions the stability of a numerical procedure such as Euler's method may also be adversely affected by an inappropriate value for $\Delta x$. Stability refers to the ability of a numerical procedure to remain viable. An application of Euler's method is shown in Figure 36. The dependent variable $y$ is fluctuating rapidly, but in Figure 36(a) the size of $\Delta x$ is relatively small and, as a result, the estimated values of y stray a bit from the true values of $y$, but the procedure remains "calm". In Figure 36(b), however, the size of $\Delta x$ is large and the numerically derived values are becoming less correct as $x$ increases. In other words, the approximations fail to keep up with changes in $y$ and the procedure appears to be out of control and getting worse with each step. From this example and the previous one, it should be evident that stormwater modelers must be careful when selecting a time step ($\Delta t$) or distance interval ($\Delta x$). This is true not only for the purposes of numerical approximation, but, as will be discussed in Chapter 3, for hydrologic and hydraulic reasons as well.

Users need to be aware of not only the way in which an urban watershed is conceptualized, but of the methods used to manipulate those conceptualizations. Obviously, errors may arise due to an inadequate model, faulty input data, or inappropriate model parameter values. However, errors are also caused by the

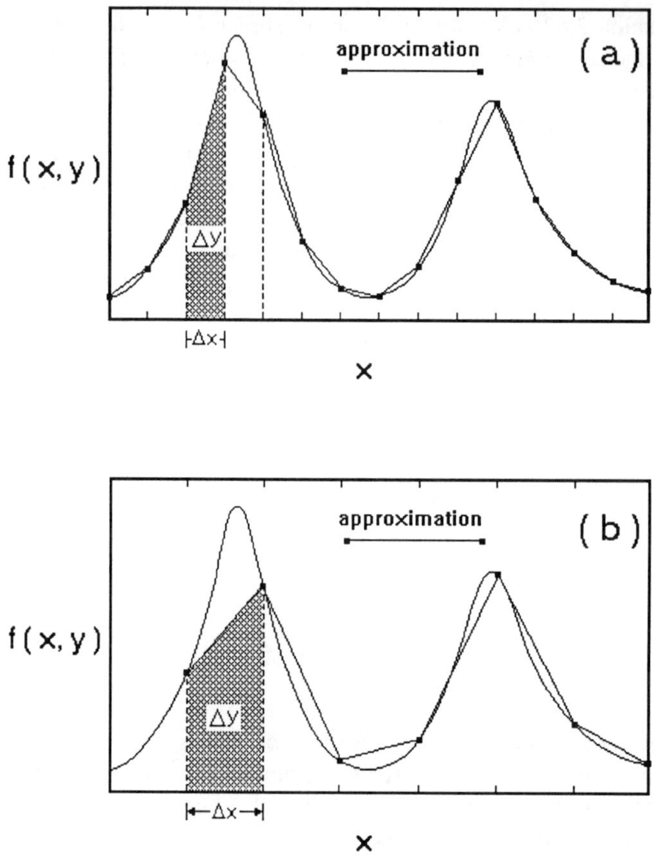


**Figure 35. Adams formula approximation of dy/dx = f(x,y).**

many numerical approximations needed to execute models on digital computers. Large errors in continuity (e.g., the model predicting more runoff than possible for the amount of rainfall) or wildly fluctuating output values are indications that the numerical methods are not working properly.

## 2.10 SUMMARY

- Modeling is accomplished in a number of different ways and the reader was introduced to some of the basic concepts and terminology.
- There are a number of models available for urban stormwater applications. They range from simple "pencil and paper" approaches to computer software packages capable of building detailed models of the urban watershed. However, even the most the sophisticated models are based on straightforward hydrologic and hydraulic con-

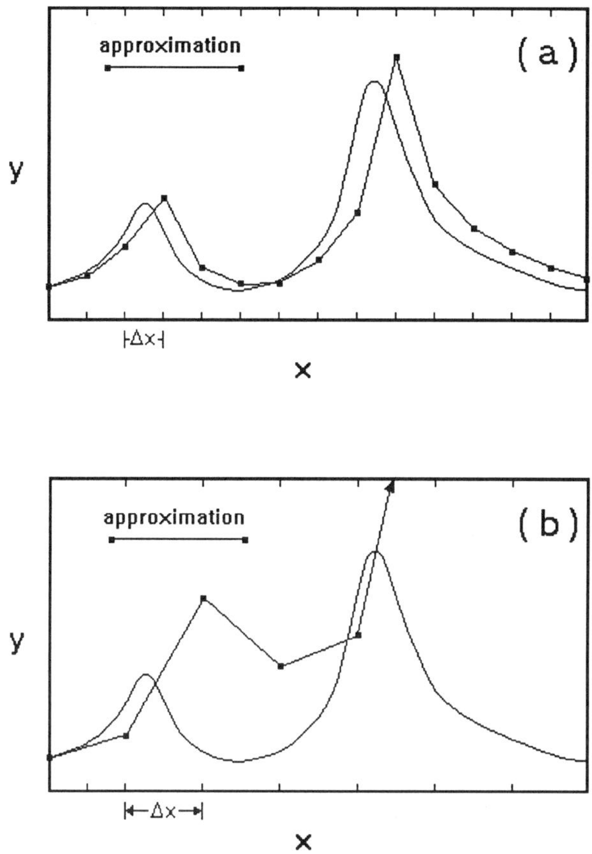


**Figure 36. Stability of Euler's method.**

cepts. All have roles to play in the analysis and solution of urban stormwater problems. The reader was introduced to three models representing the range of urban runoff modeling capabilities.

- The mathematical representations used in computer models must often be approximated and solved by numerical techniques. The computer model user must be wary of the special problems these techniques can present.

## REFERENCES

Ambrose, R.B., S.B. Vandergrift, and T.A. Wool, *WASP4, A Hydrodynamic and Water Quality Model — Model Theory, User's Manual and Programmer's Guide*, EPA-600/3-87-039, U.S. Environmental Protection Agency, Athens, GA, 1988.

American Public Works Association, *Water Pollution Aspects of Urban Runoff*, Contract WP-20-15, Federal Water Pollution Control Administration, Washington, DC, 1969.

American Public Works Association, *Urban Stormwater Management*, Special Report No. 49, APWA, Chicago, 1981.

Brandstetter, A.B., *Assessment of Mathematical Models for Storm and Combined Sewer Management*, EPA-600/2-76-175a, U.S. Environmental Protection Agency, Cincinnati, August 1976.

Chow, V.T., D.R. Maidment, and L.W. Mays, *Applied Hydrology*, McGraw-Hill, New York, 1988.

Chu, C.S. and C.E. Bowers, Computer Programs in Water Resources, WRRC Bulletin 97, Water Resources Research Center, University of Minnesota, Minneapolis, November 1977.

DiToro, D.M. and M.J. Small, "Stormwater Interception and Storage," *J. Environ. Eng. Div.*, Vol. 105, No. EE1, February 1979, pp. 43–54.

Geiger, W.F. and H.R. Dorsch, *Quantity-Quality Simulation (QQS): A Detailed Continuous Planning Model for Urban Runoff Control, Volume I, Model Description, Testing and Applications*, EPA-600/2-80-011, U.S. Environmental Protection Agency, Cincinnati, March 1980.

Hall, M.J., *Urban Hydrology*, Elsevier Applied-Science, New York, 1984.

Heaney, J.P., W.C. Huber, H. Sheikh, M.A. Medina, Jr., J.R. Doyle, W.A. Peltz, and J.E. Darling, *Urban Stormwater Management Modeling and Decision-Making*, EPA-670/2-75-022, U.S. Environmental Protection Agency, Cincinnati, May 1975.

Heaney, J.P., W.C. Huber, and S.J. Nix, *Storm Water Management Model: Level I — Preliminary Screening Procedures*, EPA-600/2-76-275, U.S. Environmental Protection Agency, Cincinnati, October 1976.

Heaney, J.P., W.C. Huber, M.A. Medina, Jr., M.P. Murphy, S.J. Nix, and S.M. Hasan, *Nationwide Assessment of Combined Sewer Overflows and Urban Stormwater Discharges: Volume II, Cost Assessment*, EPA-600/2-77-064, U.S. Environmental Protection Agency, Cincinnati, March 1977.

Heaney, J.P. and S.J. Nix, *Storm Water Management Model: Level I — Comparative Evaluation of Storage-Treatment and Other Management Practices*, EPA-600/2-77-083, U.S. Environmental Protection Agency, Cincinnati, October 1976.

Howard, C.D.D., "Theory of Storage and Treatment-Plant Overflows," *J. Environ. Eng. Div.*, Vol. 102, No. EE4, August 1976, pp. 709–722.

Huber, W.C., "Deterministic Modeling of Urban Runoff Quality," in *Urban Runoff Pollution*, Torno, H., J. Marsalek, and M. Desbordes, Eds., NATO ASI Series, Series G: Ecological Sciences, Vol. 10, Springer-Verlag, Berlin, 1986a, pp. 166–242.

Huber, W.C., "Modeling Urban Runoff Quality: State-of-the-Art," in *Proceedings of Engineering Foundation Conference on Urban Runoff Quality —*

*Impact and Quality Enhancement Technology*, Urbonas, B. and L.A. Roesner, Eds., American Society of Chemical Engineers, New York, June, 1986b, pp. 34–38.

Huber, W.C. and R.E. Dickinson, *Storm Water Management Model, Version 4, User's Manual*, EPA-600/3-88-001a, U.S. Environmental Protection Agency, Athens, GA, 1988.

Huber, W.C. and J.P. Heaney, "Operational Models for Stormwater Quality Management," in *Environmental Impact of Nonpoint Source Pollution*, Overcash, M.R. and J.M. Davidson, Eds., Ann Arbor Science, Ann Arbor, MI, 1980, pp. 397–444.

Huber, W.C. and J.P. Heaney, "Analyzing Residuals Discharge and Generation from Urban and Non-Urban Land Surfaces," in *Analyzing Natural Systems, Analysis for Regional Residuals—Environmental Quality Management*, Basta, D.J. and B.T. Bower, Eds., Resources for the Future, Washington, DC, The Johns Hopkins University Press, Baltimore, 1982, pp. 121–243. (This is also available as EPA-600/3-83-046, U.S. Environmental Protection Agency.)

Huber, W.C., J.P. Heaney, and B.A. Cunningham, *Storm Water Management Model (SWMM) Bibliography*, EPA-600/3-85-077, U.S. Environmental Protection Agency, Athens, GA, September 1985.

Huber, W.C., J.P. Heaney, M.A. Medina, Jr., W.A. Peltz, H. Sheikh, and G.F. Smith, *Storm Water Management Model User's Manual — Version II*, EPA-670/2-75-017, U.S. Environmental Protection Agency, Cincinnati, March 1975.

Huber, W.C., J.P. Heaney, S.J. Nix, R.E. Dickinson, and D.J. Polmann, *Storm Water Management Model User's Manual, Version III*, EPA-600/2-84-109a, U.S. Environmental Protection Agency, Cincinnati, 1984.

Hydrologic Engineering Center, *Storage, Treatment, Overflow, Runoff Model, STORM*, Generalized Computer Program 723-S8-L7520, Hydrologic Engineering Center, Corps of Engineers, Davis, CA, August 1977.

Hydroscience, Inc., *Areawide Assessment Procedures Manual*, EPA-600/9-76-014, U.S. Environmental Protection Agency, Cincinnati, July 1976.

Hydroscience, Inc., *A Statistical Method for the Assessment of Urban Stormwater*, EPA-440/3-79-023, U.S. Environmental Protection Agency, Washington, DC, May 1979.

Johanson, R.C., J.C. Imhoff, and H.H. Davis, *User's Manual for Hydrological Simulation Program — Fortran (HSPF): Users Manual for Release 8.0*, EPA-600/3-84-066, U.S. Environmental Protection Agency, Athens, GA, 1984.

Metcalf and Eddy, Inc., University of Florida, and Water Resources Engineers, Inc., *Storm Water Management Model, Vols. I, II, III, and IV,* EPA Reports 11024 DOC 07/71, 08/71, 09/71, 10/71, U.S. Environmental Protection Agency, Washington, DC, 1971.

Moffa, P.E., S.J. Nix, P.L. Freedman, J. Marr, and R. Field, *Control and Treatment of Combined-Sewer Overflows*, Van Nostrand Reinhold, New York, 1990.

Nix, S.J., Analysis of Storage/Release Systems in Urban Stormwater Quality Management, Ph.D. dissertation, University of Florida, Gainesville, FL, 1982.

Nix, S.J., "Applying Urban Runoff Models," *Water Environ. Technol.,* Vol. 3, No. 6, June 1991, pp. 47–49.

Nix, S.J. and J.P. Heaney, "Optimization of Stormwater Storage-Release Strategies," *Water Resources Res.*, Vol. 24, No. 11, November 1988, pp. 1831–1838.

Nix, S.J., J.P. Heaney, and W.C. Huber, "Predicting Suspended Solids Removal in Stormwater Detention Systems," *J. Environ. Eng.,* Vol. 114, No. 6, December 1988, pp. 1331–1343.

Nix, S.J., P.E. Moffa, and D.P. Davis, "The Practice of Combined Sewer System Modeling," *Water Resources Bull.*, Vol. 27, No. 2, April 1991, pp. 1–12.

Pitts, W.T., R. Drew, W.B. Heller, J. Elliott, and F.A. Eidsness, Jr., *Non-Point Source Pollution Control*, Larimer-Weld Regional Council of Governments, Loveland, CO, September 1977.

Roesner, L.A., "A Storage, Treatment, Overflow and Runoff Model for Metropolitan Master Planning," in *Short Course Proceedings: Applications of Stormwater Management Models - 1976*, F.A. DiGiano et al., Eds., EPA-600/2-77-065, U.S. Environmental Protection Agency, Cincinnati, March 1977.

Roesner, L.A., R.P. Shubinski, and J.A. Aldrich, *Storm Water Management Model User's Manual, Version III: Addendum I, EXTRAN*, EPA-600/2-84-109b, U.S. Environmental Protection Agency, Cincinnati, 1984.

Roesner, L.A., J.A. Aldrich, and R.E. Dickinson, *Storm Water Management Model, User's Manual, Version 4: Addendum I, EXTRAN*, EPA-600/3-88-001b, U.S. Environmental Protection Agency, Cincinnati, August 1988.

Soil Conservation Service, *National Engineering Handbook*, Section 4, Hydrology, Soil Conservation Service, Washington, DC, 1985.

Stankowski, S.J., *Magnitude and Frequency of Floods in New Jersey with Effects of Urbanization*, Special Report 38, U.S. Geological Survey, Water Resources Division, Trenton, NJ, 1974.

U.S. Environmental Protection Agency, *EPA Environmental Data Base and Model Directory*, Information Clearinghouse (PM 211A), U.S. Environmental Protection Agency, Washington, DC, July 1983.

Viessman, Jr., W., G.L. Lewis, and J.W. Knapp, *Introduction to Hydrology*, 3rd ed., Harper and Row, New York, 1989.

Whipple, Jr., W., N.S. Grigg, T. Grizzard, C.W. Randall, R.P. Shubinski, and L.S. Tucker, *Stormwater Management in Urbanizing Areas*, Prentice-Hall, Englewood Cliffs, NJ, 1983.

# 3 BUILDING THE MODEL OF THE URBAN WATERSHED

## 3.1 OVERVIEW

Most public and proprietary "models" are really software packages that allow the user to build a model of a watershed. The objective of this chapter is to investigate the issues involved in using the building blocks provided by these packages to construct a working urban watershed model.

Again, it is important to realize that a model is essentially a laboratory in which experiments are constructed and conducted, and, much like experiments in a laboratory, the results of "simulation experiments" must stand up to critical scrutiny. The keys to success are proper experimental setup and methodology.

## 3.2 SETTING UP THE MODEL

Before embarking on a modeling study of any kind one must carefully assess the conditions under which the study will be conducted. What is to be accomplished? Which model will be used? How will the model be used? How well must the model perform? Which watershed data are needed and which are available? Which rainfall and runoff quantity and quality data are needed and available?

### 3.2.1 What are the Modeling Objectives?

It is surprising how often modeling studies begin without a well-defined set of objectives. What serious laboratory experiment or set of experiments would be conducted without clear objectives? Why should a simulation experiment be any different?

The objectives of most urban runoff pollution modeling studies fall along the continuum illustrated in Figure 1. At one end of the spectrum are simple studies intended to estimate long-term average pollutant loads (i.e., annual loads) and develop preliminary control strategies. At the other end are studies intended for detailed analysis or design work. The detailed study will, of course, require considerably more effort and care than the simple preliminary study. To continue the laboratory analogy, determination of the concentrations of a number of complex organic compounds requires much more effort and care than running a test to determine total organic carbon (TOC). Matching the modeling effort to the study objectives is a key issue, if not *the* key issue.

### 3.2.2 Which Model Should be Used?

No model is perfect and each has its advantages and liabilities. The selection of a model is not a trivial exercise and paying it little attention can result in a poor experience.

An ill-suited model may well be worse than no model at all. A model that needs a large, detailed data base when the project budget contains neither the time nor money to provide that data will prove to be unsatisfactory. On the other hand, a model designed to provide rough estimates will be of little use in design work. There is much to consider.

Figure 2 presents the problem in terms of benefits and costs. In general, the more complex a model, the more information it is capable of producing. However, the cost of that information is rising rapidly. What are the benefits of that information? More information is good, but at the margin each new piece provides less and less additional value. This means, of course, that there is an ideal point at which we can maximize the net gain or benefits (shown in Figure 2). In other words, the best model maximizes the net gain.

This is all rather academic, of course. Benefits and costs (benefits in particular) are difficult to assess. However, the basic message is obvious — a model should be selected carefully. Most importantly, a model should meet the technical goals of a particular application. But other, more intangible, factors are also important (Moffa et al., 1990):

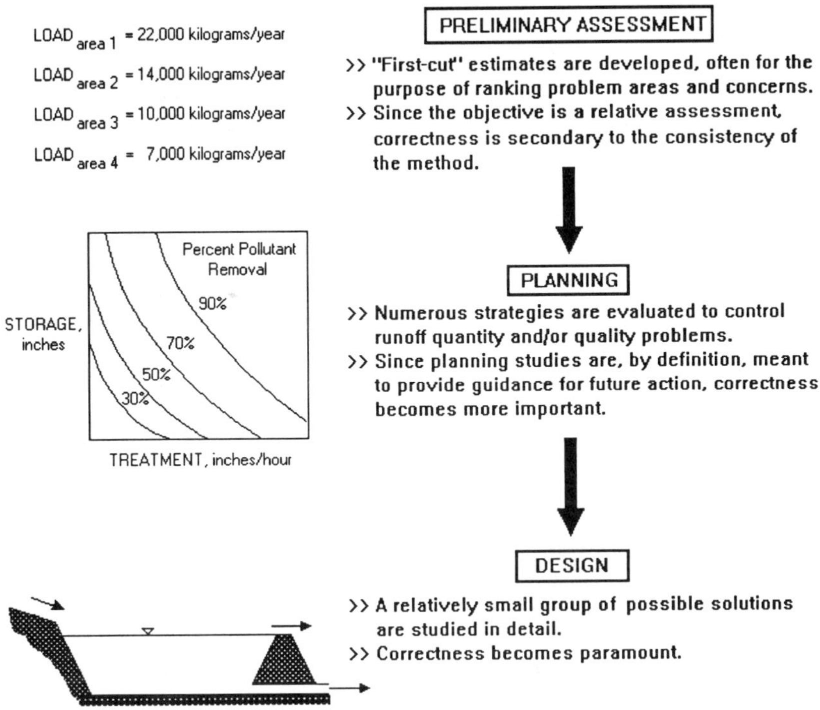


**Figure 1. Continuum of urban runoff modeling objectives.**

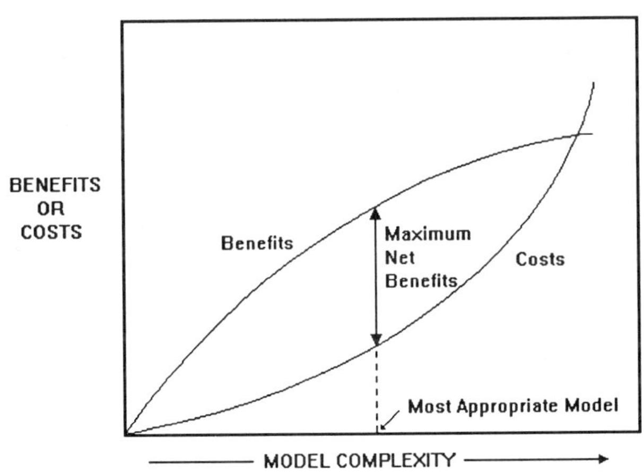


**Figure 2. Benefits and costs as a function of model complexity.**

- *Availability of suitable hardware.* The personal computer (and its ever-increasing capability) has given analysts and engineers the computing resources needed to operate a wide array of models. Even very large models can be executed on personal computers, but they may require more sophisticated processors and other special hardware.
- *Availability of trained personnel.* Urban runoff models are not known for their user friendliness and even if they were it takes a fairly good water resource or environmental engineer to use them properly. In other words, the user must be competent enough to critically evaluate the model output.
- *Long-term commitment to the model.* It may be desirable to develop expertise with a particular model even if that model is not ideal for the current application. If a number of applications are anticipated, it may be more beneficial to invest heavily in one model than to switch models from project to project.
- *In-house model experience.* Occasionally, in-house experience with a particular model is readily available. In this case, the fact that no training or "warm-up" period is necessary may overwhelm some of the deficiencies of the model.
- *Acceptance and support of the model.* It is possible that a fairly obscure model will meet the needs of a given project or series of projects. However, there is no substitute for learning from the experiences of colleagues. If a model is used by few it becomes that much more difficult to establish its credibility and to interpret its output. Likewise, if there is little or no support from the model developers (i.e., to provide updates, correct errors, provide user assistance, etc.), the chances for success are diminished. It is beneficial to use a model that is well supported and has a long, well-documented history of use.
- *Commitment to modeling as a tool.* How deeply committed is the user to modeling as a tool? A user that considers modeling an integral part of a problem-solving repertoire will view individual models differently than a user that views modeling as a rare, esoteric exercise. Actually, this level of commitment is strongly connected to several of the factors discussed above.

### 3.2.3 How Much Physical Detail is Needed?

Essentially all urban runoff models are "lumped" models, i.e., the physical characteristics of the watershed are assumed to be spatially constant. This is in contrast to more sophisticated models that directly account for the physical heterogeneities in the watershed (i.e., distributed models). A lumped model can be made to behave somewhat like a distributed model if the watershed is divided into several "subwatersheds". In this case, each subwatershed is given spatially constant characteristics, but together the subwatersheds form a distribution of characteristics. The spatial detail of a watershed can be increased by dividing it into more subwatersheds. This is known as *discretization.*

In general heterogeneous watersheds require more highly discretized models than relatively homogeneous watersheds. However, additional detail that adds little or nothing in terms of new information is wasteful (recall Figure 2). The same can be said for the discretization of a sewer system. However, the appropriate level of discretization or detail is really most strongly tied to the modeling objectives. For example, if changes in a number of smaller trunk sewers are anticipated as part of a control plan, the user would be wise to include that part of the sewer system in the model. Likewise, if heavy development is predicted for a particular watershed, it would be wise to use several subwatersheds, so that the changes in model parameters are more straightforward.

The appropriate level of discretization is also related to how the model responds to different levels of discretization. An example is useful. Figure 3 demonstrates the response of the Storm Water Management Model (see Chapter 2) to various discretization schemes applied to a watershed in Baltimore, Maryland (Metcalf and Eddy, Inc. et al., 1971). One can see that the runoff hydrographs produced by the 12-subwatershed and 5-subwatershed schemes are not very different. In this case, little seems to be gained by dividing the watershed into 12 subwatersheds (rather than just 5). This is actually an interesting application of sensitivity analysis in which the effect of too little spatial detail is assessed. A discussion of sensitivity analysis can be found in Section 3.4.

### 3.2.4 How Finely Should Time Be Divided?

Most models rely on numerical methods to approximate and solve one or more differential and/or integral equations (see discussion and examples in Chapter 2). These methods often divide time into finite intervals, i.e., time steps. Some models fix the time step, while others allow the user to select it. In the latter case, the user should heed the guidelines provided by the model developers, but the effect of different time steps should probably be examined.

As pointed out in Chapter 2, the time-step value affects two things: (1) the ability of the model to keep apace of the response time of the system (i.e., watershed) and (2) the stability of the numerical solution procedures. Large time steps are generally more problematic than small time steps, but how is the appropriate time step selected? An easy test to conduct gradually shortens the

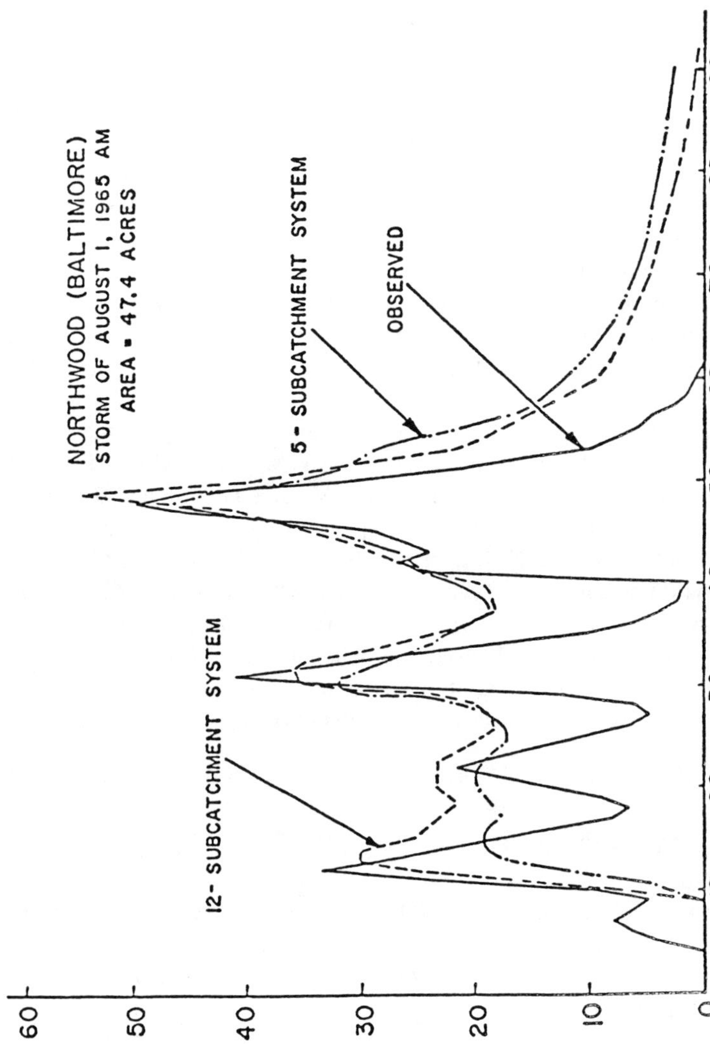


**Figure 3. Response of SWMM to different spatial discretization levels (from Metcalf and Eddy, Inc. et al., 1971).**

time step over a sequence of test runs until the output of interest (e.g., runoff volume) no longer changes significantly. At this point the appropriate time step has been located since it is no longer causing appreciable error. Of course, smaller time steps lead directly to longer simulation run times and some compromises might be in order.

### 3.2.5 How Well Must the Model Perform?

The standard of model performance, it might be said, is in the eye of the beholder. Simply put, a useful model is one capable of performing at an *acceptable* level of competence. There are no arbitrary standards for model performance nor is there a need for any. The user is ultimately responsible for his or her work and, thus, must be satisfied that the model's performance meets the needs of the study. The criteria used to assess performance are another matter. Some of these issues will be explored in Section 3.3.2.

### 3.2.6 Should Single-Event or Continuous Simulation Be Used?

The issue of whether a modeling study should focus on a single "design" event or on a long continuous series of events is not as simple as it might seem. In general, continuous simulation is better suited for planning purposes and preliminary assessments, whereas single-event simulation is better suited for the analysis of design conditions. This distinction is beginning to blur, however, especially as computer speed increases. Ultimately, this is a question that must be addressed in the context of the application, but some of the following observations may be useful (Moffa et al., 1990; Nix, 1991):

- Model parameters adjusted through a calibration exercise (see Section 3.3.2) for single-event simulations may not be applicable for continuous simulation. Parameters adjusted in this fashion are affected by the antecedent watershed conditions (i.e., soil moisture, water in depression storage, etc. at the start of the simulation run) assumed for each storm event. In addition, it is often necessary to use a longer time step in continuous simulation, and results are often influenced by the time-step size. Calibration of a model intended for continuous simulation should be conducted with continuous model runs.
- The use of a "design" rainstorm (e.g., 10-year storm) is fundamentally flawed because the resulting runoff event does not necessarily have the same return period. This is principally due to the fact that antecedent conditions have a great deal of influence on the severity

of a particular storm. For example, a 2-year rainstorm may result in a 5- or 10-year (or larger) runoff event if the ground is already at or near saturation. Continuous simulation can produce the response of a watershed to many years of rainfall. From these responses, the user can identify critical *runoff* events that may be suitable for "design" event simulations (Bedient and Huber, 1992).

- Continuous simulation lessens the importance of estimating the antecedent condition of the watershed. In single-event simulation, the initial condition of the catchment can significantly affect the model output (and, of course, the actual watershed). A long, continuous simulation (e.g., a year) would only be affected for the first few events and the overall impact would be minimal.

## 3.3 ESTABLISHING MODEL CREDIBILITY

All urban runoff models are imperfect representations of the urban watershed. In addition to structural misrepresentations, errors associated with model parameters and input data can cause models to poorly depict watershed behavior. Therefore, in most applications a model must be massaged into a working, useful tool. (The exceptions might be cases in which a model is being used for conceptual or preliminary planning. In these cases, the model is being used for initial guidance and its ability to accurately mimic watershed behavior is not as important.) Further complicating the matter is the fact that data used to characterize "true" watershed behavior (e.g., runoff flow or pollutant measurements) are themselves flawed.

A model proves its worthiness in two ways. First, does it reproduce the behavior of the urban watershed accurately? Second, does it do so with precision? *Accuracy* is the ability of the model to replicate the true system. *Precision*, the more difficult concept, has to do with reproducibility. Perhaps a better way of looking at the two ideas is to look at their opposites. Inaccuracy is related to systematic bias. Imprecision is strongly tied to uncertainty or random errors.

An analogy is useful at this point. Take four archers each with ten arrows. Archer A fires her ten arrows as shown on target A in Figure 4. Archers B, C, and D do likewise with targets B, C, and D, respectively. Clearly archer A would win any competition among the four. Not only is she accurate, but she is also quite precise. Archer B, on the other hand, seems to be precise, but off the bull's eye (i.e., inaccurate). Archer C scatters his arrows around the bull's eye; imprecision, not accuracy seems to be his problem. Archer D seems to lack accuracy and precision with her arrows not only well off the bull's eye but scattered as well. If you were the coach or instructor assigned the duty of improving the skills of archers B, C, and D, which would be the easiest with which to work? Most likely, it would be archer B. Chances are good that the

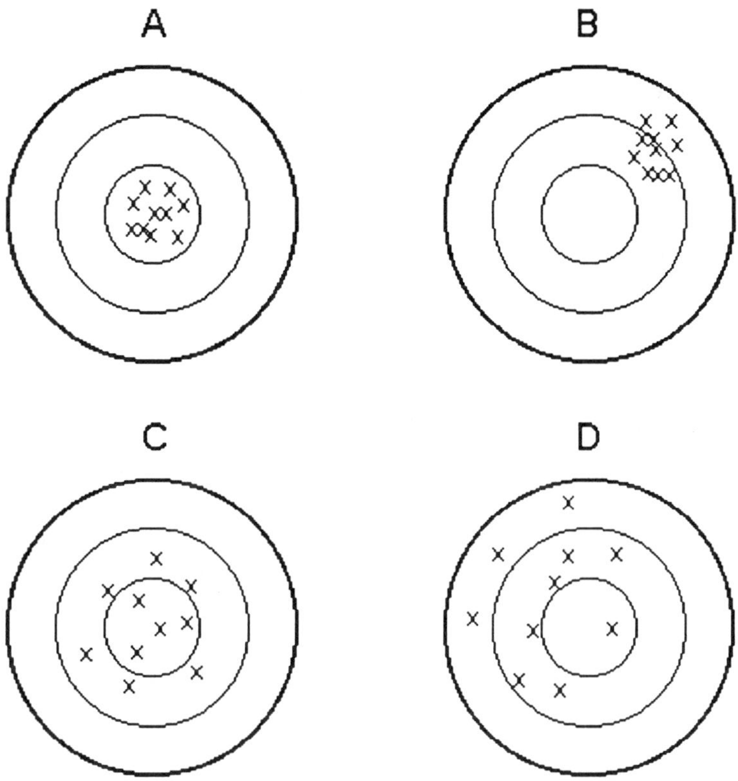


**Figure 4. Accuracy and precision in archery.**

inaccuracy can be removed by eliminating some sort of systematic bias such as a poorly adjusted sight. Archer C seems to have no problems with systematic bias, but uncertainties or inconsistencies seem to plague him. These would be far more difficult to remove. Archer D may be hopeless.

What does this little story tell us in terms of urban stormwater modeling? A model's errors can also be divided into those associated with systematic error and random error. Systematic errors are easier to eliminate because of their very nature (i.e., their "systematic" nature). Random errors are difficult if not impossible to eliminate because we understand them so poorly. Systematic error and random error actually form two ends of a continuum where the former is error we understand and the latter is error we do not understand. Generally, the more we understand about a system the less important random error becomes. When establishing the credibility of a model the user must deal with both kinds of error. Systematic error is reduced through the process of validation. Random error or uncertainty is difficult to reduce, but at the very least random error and its effects must be assessed so that the model results can be properly interpreted. This can be done with sensitivity analysis.

McCuen and Snyder (1986) present rigorous (and excellent) discussions on many of the topics found here. The discussion here is much simpler and the interested reader may want to refer to their text.

### 3.3.1 Verification

A useful precursor to validation and sensitivity analysis is the often-overlooked process of verification. This is unfortunate since much can be learned about a model's behavior with a minimal investment of effort.

*Verification* involves a "check out" of a model's components to see if they perform as expected and to learn the model's idiosyncrasies. This can often be done with sample data and output provided by the model developer. Verification builds user confidence in the model, helps the user understand the nuances of the model, and makes the later modeling efforts more efficient.

A simple example is shown in Figure 5. Here the model is asked to simulate the response of a watershed to steady rainfall. In this case, the model's response seems to be reasonable and what is expected. The key to successful verification is the use of situations for which the outcome is well known.

### 3.3.2 Validation

*Validation* is the process of collecting data to describe the behavior and characteristics of the urban watershed (i.e., precipitation, runoff quantity, runoff quality, etc.) for a wide range of conditions *and* adjusting the model so that it adequately replicates the watershed as depicted by these data. Validation is probably the most important process in establishing model credibility. It is summarized in Figure 6 and described in more detail below.

"Validation" is not the term that most stormwater modelers prefer for this process. However, it is a descriptive term and is consistent with the terminology used in systems science. Many refer to this process as *calibration and verification*. This phrase refers to a process in which the model parameters are adjusted so that the model agrees with, or is calibrated for, one set of system data (e.g., a rainfall event). The "calibrated" model is then "verified" against an independent set of data (e.g., another rainfall event). This is probably poor terminology. First, the process of adjusting model parameters so that the model adequately simulates the system is actually an iterative process involving a lot of "calibration" and "verification" with several independent sets of data. It seems preferable to simply call this part of the validation process "calibration" and to use the term "verification" for the check out of the model algorithms (see earlier discussion). Second, the validation process includes the important tasks of calibration *and* data collection. This more encompassing term, i.e., validation, is used here to help make it clear that these two activities are closely intertwined, with each helping to guide the other.

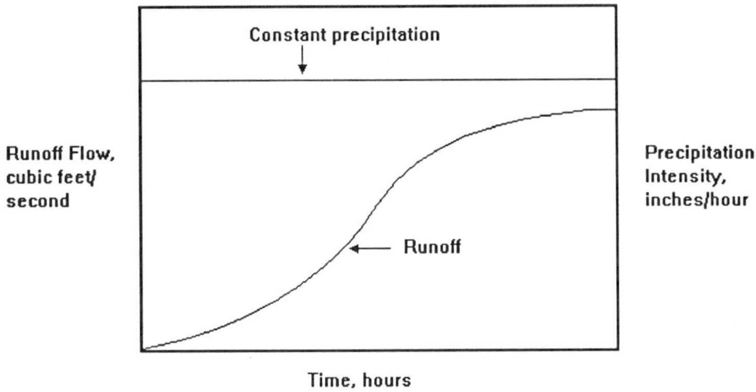


**Figure 5. Response of a watershed to a steady rain event.**

STEP 1 - PRIORITIZE THE MODEL FACTORS

>> Determine which parts of the model are pertinent to the application.

>> Use sensitivity analysis to determine which model factors ( parameters, variables, etc. ) have a significant impact on model output.

STEP 2 - CLASSIFY THE IMPORTANT MODEL FACTORS

>> Determine which of the "important" model factors ( as determined by Step 1 ) are measurable and have a clear physical meaning.

>> Those factors that do not meet the above criteria will most likely be used to calibrate the model.

STEP 3 - DESIGN AND IMPLEMENT THE DATA COLLECTION PROGRAM

>> Design a data collection program that meets the overall study objectives. Be sure that the data collection and modeling efforts are complementary.

>> The data used for calibration should cover the range of conditions under which the model will be used to make predictions.

STEP 4 - CALIBRATE THE MODEL

>> Adjust the parameters that were identified in Step 2 so that the simulated model output agrees with the actual measurements made during several independent events or periods.

>> The calibration process is iterative, with adjustments made between several independent events or periods. The end result will probably be a compromise with no one event calibrated perfectly, but many "satisfactorily".

**Figure 6. Steps in model validation.**

The steps involved in validation are described below and an example is used throughout the discussion. The example involves a small urban watershed used in a report by Brandstetter (1976) to conduct a comparison of several models, including the Storm Water Management Model (SWMM). The watershed, called the Oakdale Avenue catchment, is located in Chicago and has an area of 12.9 acres (5.22 ha). The watershed is entirely residential and is shown in Figure 7.

The Oakdale Avenue watershed is drained by a 30-in. (76-mm) reinforced concrete sewer. This sewer in turn drains to a 10.5 ft × 10.5 ft (3.20 m × 3.20 m) trunk sewer. The watershed was divided into 13 subwatersheds, each draining to an inlet manhole. (The number of subwatersheds is probably excessive, but these are the data provided by Brandstetter.) The subwatersheds are shown in Figure 8 along with the drainage system. Table 1 lists the physical characteristics of each subwatershed. Table 2 gives the soil drainage characteristics assumed for the entire watershed. Table 3 displays the physical characteristics of the 14 conduits making up the drainage system.

The watershed characteristics shown in Tables 1, 2, and 3 were used by Brandstetter to test the original versions of the RUNOFF and TRANSPORT Blocks of SWMM (Metcalf and Eddy, Inc. et al., 1971), among other models. Here, a recent version of the RUNOFF Block (Huber and Dickinson, 1988) was used to simulate the surface flows and the flows through the drainage system. The TRANSPORT Block was not used here in order to keep the application as simple as possible. The RUNOFF Block has a simple, but seldom used, gutter/conduit routing routine. Pollutants were not simulated. Version 4 of SWMM was used to carry out the calculations.

*Step 1 — Prioritize the Model Factors.* A typical model requires the user to consider and quantify several input variables (e.g., rainfall), initial values for state variables (e.g., initial pollutant load on the surface), and parameters (e.g., imperviousness). As a group these will be called "model factors". Some factors have direct impact on model output; some have minimal impact. For example, the percent of the watershed covered with impervious surfaces will probably have an important effect on runoff volume. On the other hand, parameters associated with infiltration are not very important when dealing with a mostly impervious watershed. As another example, consider evaporation (often an input variable). In cases where short term (perhaps only a single runoff event) simulation is the objective, the impact of evaporation is probably very small. It is important for the user to identify the important and unimportant model factors so that resources are used wisely.

An application of *sensitivity analysis* can be useful here. This process systematically varies model factors to determine which have an important impact on pertinent facets of model output (pertinent to the modeling objectives). At this phase of the process calibrated parameter values are not available, but reasonable first estimates can be used as a reference point. Remember, all this step is trying to accomplish is the identification of the factors to which the model is most sensitive.

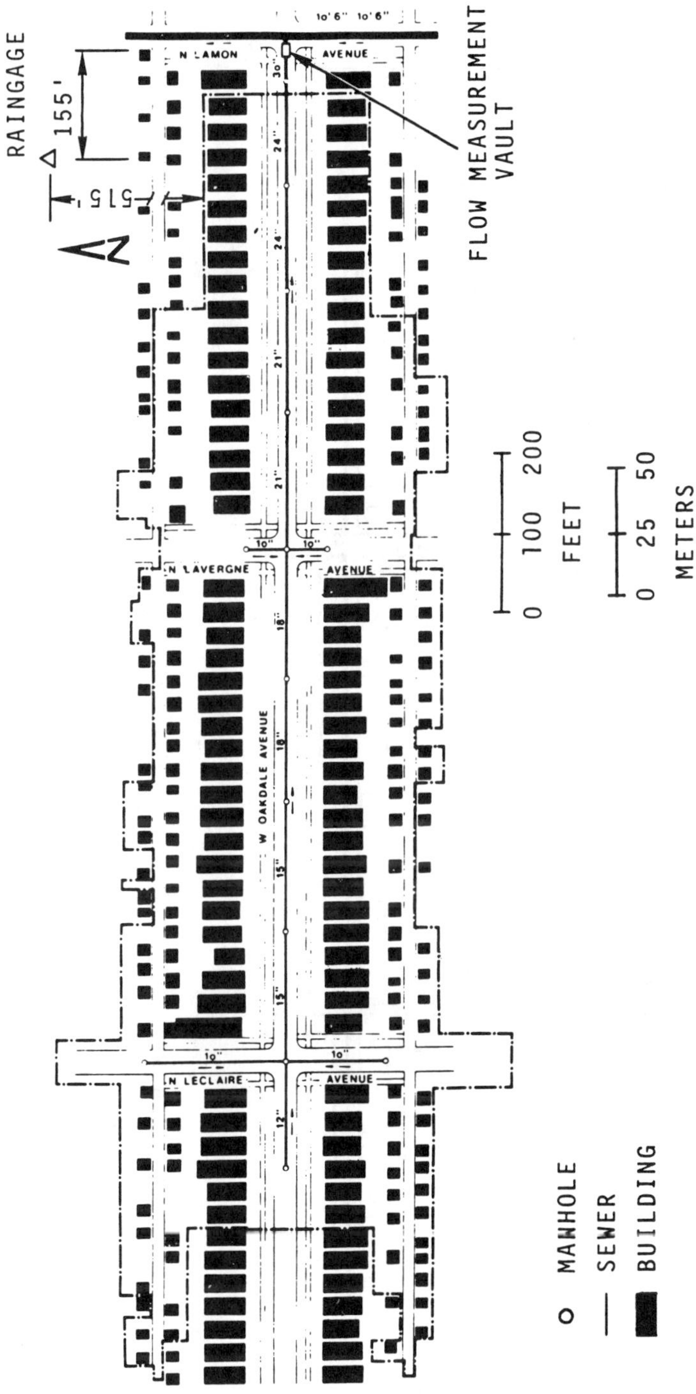


**Figure 7.** **Oakdale Avenue catchment, Illinois (from Brandstetter, 1976).**

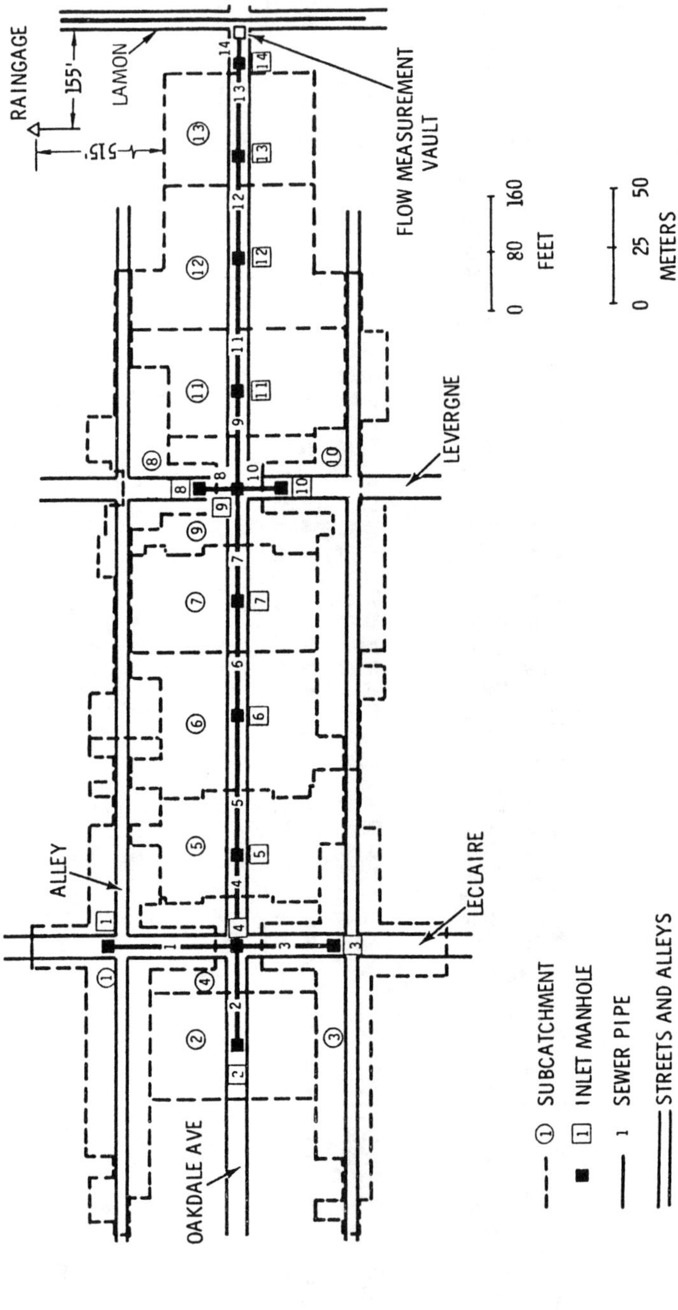


**Figure 8. Subwatersheds in Oakdale Avenue catchment (from Brandstetter, 1976).**

Table 1. Physical Characteristics of Oakdale Avenue Catchment (from Brandstetter, 1976)

| Subcatchment number | Inlet manhole number | Length | | Width | | Drainage area | | Imperviousness (%) | Ground slope (%) | Manning's n | |
|---|---|---|---|---|---|---|---|---|---|---|---|
| | | m | ft | m | ft | ha | acres | | | Impervious area | Pervious area |
| 1 | 1 | 212.4 | 697 | 30.5 | 100 | 0.65 | 1.60 | 44.0 | 0.60 | 0.013 | 0.350 |
| 2 | 2 | 74.4 | 244 | 45.7 | 150 | 0.34 | 0.84 | 44.0 | 0.90 | 0.012 | 0.350 |
| 3 | 3 | 212.4 | 697 | 24.4 | 80 | 0.52 | 1.28 | 51.7 | 0.60 | 0.013 | 0.350 |
| 4 | 4 | 166.1 | 545 | 12.2 | 40 | 0.20 | 0.50 | 55.4 | 0.90 | 0.013 | 0.350 |
| 5 | 5 | 78.9 | 259 | 48.8 | 160 | 0.38 | 0.95 | 39.5 | 0.90 | 0.012 | 0.350 |
| 6 | 6 | 78.3 | 257 | 61.0 | 200 | 0.48 | 1.18 | 41.7 | 0.90 | 0.012 | 0.350 |
| 7 | 7 | 75.6 | 248 | 48.8 | 160 | 0.37 | 0.91 | 40.5 | 0.90 | 0.012 | 0.350 |
| 8 | 8 | 147.8 | 485 | 24.4 | 80 | 0.36 | 0.90 | 54.0 | 0.37 | 0.013 | 0.350 |
| 9 | 9 | 195.0 | 643 | 12.2 | 40 | 0.24 | 0.59 | 56.5 | 0.90 | 0.013 | 0.350 |
| 10 | 10 | 174.0 | 571 | 27.4 | 90 | 0.48 | 1.18 | 47.4 | 0.37 | 0.013 | 0.350 |
| 11 | 11 | 78.9 | 259 | 48.8 | 160 | 0.38 | 0.95 | 40.6 | 0.90 | 0.012 | 0.350 |
| 12 | 12 | 80.5 | 264 | 61.0 | 200 | 0.49 | 1.21 | 40.6 | 0.90 | 0.012 | 0.350 |
| 13 | 13 | 68.0 | 223 | 48.8 | 160 | 0.33 | 0.82 | 43.1 | 0.90 | 0.012 | 0.350 |

*Note:* Length is in the direction of flow and width is perpendicular to the direction of flow, for a rectangular area of a size equal to the actual subcatchment area.

Table 2. Soil Drainage Characteristics of the Oakdale Avenue Catchment (from Brandstetter, 1976)

| Item | Units | |
|---|---|---|
| | Metric | British |
| Pervious areas | | |
| Detention storage capacity | 5.1 mm | 0.20 in. |
| Maximum infiltration rate | 63.5 mm/hr | 2.50 in./hr |
| Minimum infiltration rate | 11.4 mm/hr | 0.45 in./hr |
| Infiltration decay rate | 0.00115 $sec^{-1}$ | 0.00115 $sec^{-1}$ |
| Impervious areas | | |
| Detention storage capacity | 2.0 mm | 0.08 in. |
| Maximum infiltration rate | 0.0 mm/hr | 0.00 in./hr |
| Minimum infiltration rate | 0.0 mm/hr | 0.00 in./hr |
| Infiltration decay rate | 0.0 $sec^{-1}$ | 0.00 $sec^{-1}$ |

*Note:* The infiltration rates are defined by Horton's infiltration equation. The detention storage capacities must be satisfied before surface runoff begins.

Table 3. Characteristics of the Oakdale Avenue Drainage System (from Brandstetter, 1976)

| Pipe number | Downstream junction number | Diameter | | Length | | Invert slope, percent | Manning's n |
|---|---|---|---|---|---|---|---|
| | | m | ft | m | ft | | |
| 1 | 4 | 0.25 | 0.83 | 51.8 | 170 | 0.71 | 0.012 |
| 2 | 4 | 0.30 | 1.00 | 39.6 | 130 | 0.72 | 0.012 |
| 3 | 4 | 0.25 | 0.83 | 32.0 | 105 | 1.08 | 0.012 |
| 4 | 5 | 0.38 | 1.25 | 48.8 | 160 | 0.45 | 0.012 |
| 5 | 6 | 0.38 | 1.25 | 53.3 | 175 | 0.45 | 0.012 |
| 6 | 7 | 0.46 | 1.50 | 48.5 | 159 | 0.40 | 0.012 |
| 7 | 9 | 0.46 | 1.50 | 53.0 | 174 | 0.40 | 0.012 |
| 8 | 9 | 0.25 | 0.83 | 15.2 | 50 | 3.78 | 0.012 |
| 9 | 11 | 0.53 | 1.75 | 50.6 | 166 | 0.35 | 0.012 |
| 10 | 9 | 0.25 | 0.83 | 13.7 | 45 | 4.20 | 0.012 |
| 11 | 12 | 0.53 | 1.75 | 51.8 | 170 | 0.35 | 0.012 |
| 12 | 13 | 0.61 | 2.00 | 41.8 | 137 | 0.30 | 0.012 |
| 13 | 14 | 0.61 | 2.00 | 40.2 | 132 | 0.30 | 0.012 |
| 14 | — | 0.76 | 2.50 | 13.4 | 44 | 0.30 | 0.012 |

The results of a sensitivity analysis for the SWMM parameters involved in the Oakdale Avenue example are shown in Figures 9 and 10. The response of the watershed to a single storm was simulated with the watershed parameters (imperviousness, depression storage, etc.) set initially at the values estimated by Brandstetter. This was followed by a series of model runs in which a range of values was tested for each parameter. Each line depicts the response of the model to a percent change in the initial value of one parameter while holding all others constant. (In other words, the center point in both figures represents the parameter values estimated by Brandstetter and the associated model results.) Figure 8 shows the response in terms of the runoff volume; whereas Figure 9 shows the response in terms of the peak flow. From this simple analysis it is easy to see that runoff volume, as simulated by the SWMM RUNOFF Block, is fairly sensitive to changes in imperviousness (i.e., the percent of the watershed covered by impervious surfaces). It also appears that the peak runoff flow is sensitive to imperviousness, pipe (conduit) roughness, and the width of overland flow. Changes in the infiltration parameters have a negligible effect on both the volume and peak flow — not surprising for a very highly developed area. Other parameters also have minimal effects on the output.

The storm used in this example is an actual storm that occurred on May 19, 1959 (the details of which may be found in Brandstetter, 1976). This same storm is used in the validation effort discussed later in this section. It is not necessary to use an actual storm in this step, but whatever is used should be typical of the anticipated application.

The results of the sensitivity analysis described above should lead the user to focus on those parameters that have a significant effect on model output. In this case, it is apparent that imperviousness, pipe roughness, and width of overland flow require attention, either through accurate characterization or through calibration. On the other hand, less attention is required for the infiltration parameters, depression storage, or surface roughness, among others.

The results of sensitivity analysis can be enormously useful. Several model factors will prove to be superfluous; others might be surprisingly important. With this information, the subsequent effort can be significantly reduced.

*Step 2 — Classify the Model Factors*. It is helpful to classify the important model factors ("important" meaning that they have a significant impact on the model output) on the basis of their measurability. Those which are more conceptual (e.g., width of overland flow) tend to be difficult to measure. These are in contrast to those having clear physical meaning (watershed area, imperviousness, etc.) and relatively easy to measure. While the more conceptual parameters should be specified in a considered and consistent manner, it is unwise to spend much effort to establish *a priori* values. In fact, the most abstract parameters are usually the first to be changed when calibrating the model to actual observations. (Some guidance for initial estimates can often be obtained by a review of the literature for applications of the same model to other watersheds.) The parameters that are more directly tied to measurable quantities should be carefully estimated and then, as much as possible, left alone through the calibration process.

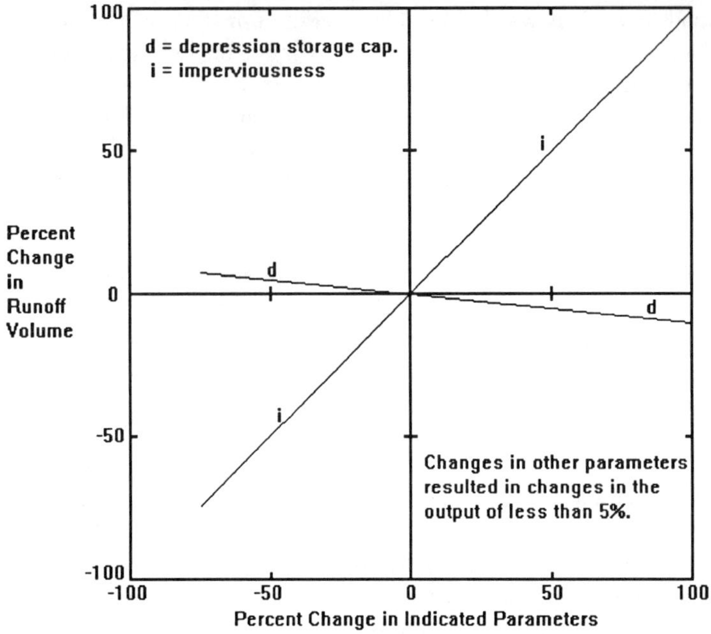


**Figure 9. Pre-calibration sensitivity analysis of Oakdale Avenue catchment model, runoff volume.**

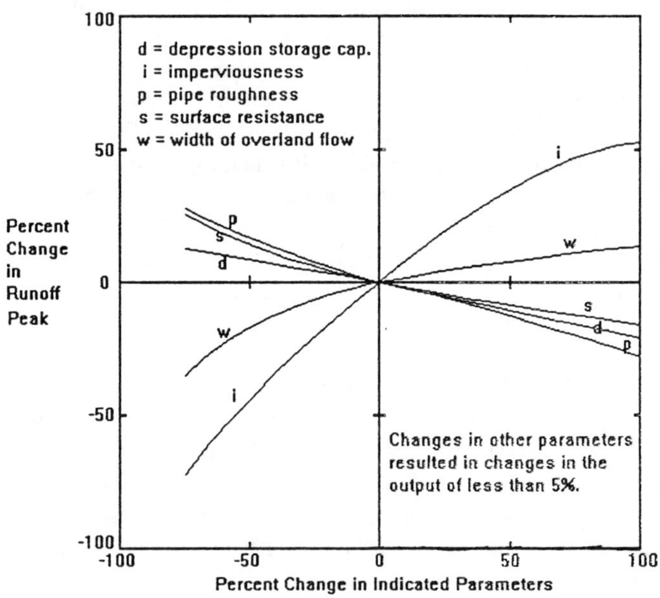


**Figure 10. Pre-calibration sensitivity analysis of Oakdale Avenue catchment model, runoff peak.**

The "width of overland flow" is probably the most conceptual and least measurable of the important parameters identified for the Oakdale Avenue watershed. The pipe roughness is a little less conceptual and a bit easier to estimate. Imperviousness is a fairly well defined physical parameter that seems straightforward to measure, although it is not always clear what is impervious and what is pervious. For example, rooftops draining to lawns should probably not be considered impervious. There is room for uncertainty in all of these parameters, but the width of overland flow should probably be adjusted the most vigorously in the calibration process and not be the object of an extensive specification effort.

*Step 3 — Design and Implement the Data Collection Program.* A useful data base of system measurements (both inputs and outputs) with which to calibrate the model can be difficult to obtain. There are a number of important issues, and all of them affect the cost and quality of the data. How many storm events should be monitored? How many rain gages are needed, and where should they be placed? At which locations should flow and quality be monitored? How often should measurements being taken? The answers to these questions can only be provided by a careful review of the nature of the watershed (e.g., the "response" time of the watershed), the modeling objectives, and the available resources (including time, money, personnel, and previously collected data). Again, it is crucial that the data collection and modeling efforts complement each other.

The collection of data to estimate model parameters and initial state variable values is also important. As mentioned earlier, this effort should focus on the important, physically measurable factors and not the more conceptual factors that are difficult to relate to measurable quantities.

The amount of data needed for calibration is a function, as always, of the modeling objectives. For single-event simulation, at least three events are probably needed — six or more would be more convincing. Using three events gives some chance for uncovering the "derivative", i.e., the trend of the trend. For continuous simulation, two or more independent periods (of several months or more, if possible) are probably adequate. More provides an opportunity to develop a bit more credibility, but the cost and time requirements may be prohibitive.

The calibration data should cover as wide a range of conditions as practical within the bounds of the study. It is likely that a model will respond to different conditions in a somewhat inconsistent manner. For example, a model "calibrated" for a small event may perform poorly for a large event and vice versa. However, the user has to be concerned with more than catching big and small events. Equally important is the capture of critical conditions. If, for example, summer thunderstorms create the largest problems, whether quantity or quality related, summer periods should be used for calibration. If the spring thaw is critical, spring periods are perhaps more appropriate.

In addition to the issues raised above, a well-executed study must be cognizant of how the model results will be employed and even some idea of what the results will be before the modeling is completed. For example, if preliminary analysis of an urban stormwater problem strongly points to control measures at a few locations, the data collection program would certainly want to focus on those areas. This warning is probably beginning to sound like a broken record, but it so important. Every modeling study should have some idea of where it will lead. Some surprises will arise, but every experiment begins with a hypothesis (which is nothing more than a statement of expected results).

*Step 4 — Calibrate the Model.* The following represents a general three-part prescription for calibrating a model:

1. Adjust the runoff "volume" parameters. It is probably better to first adjust those parameters that have the most effect on runoff volume. Without a fairly accurate representation of the runoff volumes, the process of adjusting parameters that most affect peak flows and hydrograph shapes will be much more difficult.
2. Adjust the hydrograph "peak and shape" parameters (assuming that hydrograph simulation is desired). The peak and shape of the runoff event or events should be adjusted after the simulated runoff volumes agree reasonably well with the field data. However, the effect of changing the value of one of these parameters may alter the runoff volume results to some degree and other parameters may have to be readjusted.
3. Adjust the water quality parameters. A model that is not calibrated for quantity cannot produce reliable quality results. Therefore, parameters responsible for generating pollutant loads and their transport through the watershed should be adjusted after calibrating the quantity parameters.

Recall that the focus throughout calibration should be on the more abstract model parameters that have a significant impact on model output. In other words, try to first adjust the important parameters about which there is the most uncertainty.

Calibration is an iterative process. For example, while calibrating a model to one event (or continuous sequence, if continuous simulation is the objective), the adjustment of "peak and shape" parameters will likely require readjustment of the volume parameters. "Calibrated" parameters for a second event (or second continuous sequence) will probably not agree with those of the first event. All of this inevitably leads to a compromise in which the final parameter values produce model results that probably fit no one event extremely well, but yet they allow the model to fit all or most of the measured data *reasonably* well.

What defines a "reasonable" fit or a "reasonably well calibrated" model? This is difficult to answer precisely, but easy to answer conceptually. The model is calibrated when it reproduces the system or watershed behavior well enough to meet the modeling objectives. Perhaps the more important question is "What is reasonable to fit?" In other words, by what criteria will the user judge the competence of the model? This question is also related to modeling objectives. For example, if the interest is in runoff volume alone, how close the model reproduces actual runoff volume would be the most important.

There are a number of "goodness-of-fit" criteria that can be considered (Shannon, 1975, Fleming, 1975). Figure 11 illustrates the nomenclature used in each of the criteria described below.

One very simple method minimizes the sum of the squares of the differences between predicted and observed data:

$$\text{Minimize } F = \sum_{i=1}^{n} (q_i - r_i)^2 \qquad \textbf{(3.1)}$$

where $F$ = goodness-of-fit criterion
$q_i$ = observed value $i$
$r_i$ = concurrent simulated value $i$
$n$ = number of data pairs

Another method minimizes the integral of the modulus, i.e.,

$$\text{Minimize } F = \sum_{i=1}^{n} |q_i - r_i| \qquad \textbf{(3.2)}$$

The absolute value sign ensures that positive and negative differences do not cancel each other and give a false appearance of agreement. A third method minimizes the sum of the squares of the differences between the logarithms of the observed and predicted data:

$$\text{Minimize } F = \sum_{i=1}^{n} [\ln(q_i) - \ln(r_i)]^2 \qquad \textbf{(3.3)}$$

The advantage of this measure is that it lessens the impact of deviations occurring at larger values of $q_i$ and $r_i$ (i.e., even though those deviations may be large they may be no larger on a percentage basis than deviations occurring at smaller values of $q_i$ and $r_i$).

Regardless of the criteria used to measure model performance, the question of *what* constitutes a good fit, or the standard of performance, persists. It is highly unlikely that the *best* fit will be reached in a practical modeling study

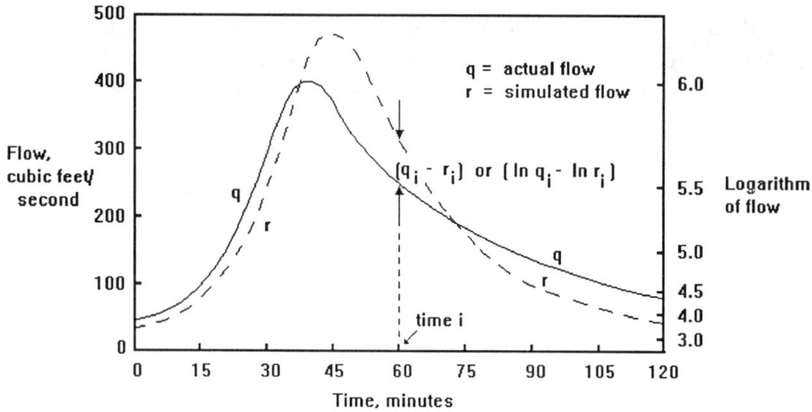


**Figure 11. Goodness of fit for model results.**

since resources are always limited. McCuen and Snyder (1986) state that the perceived success of the iterative calibration process depends on the user's "subjective assessment of whether the lack of total fit will cause errors in design." Essentially, it all comes down, again, to an assessment of costs vs. benefits and the modeling objectives (there it is again!). If significant additional effort will produce only minimal gain, what is the point of "improving" the calibration?

The kind of calibration exercise described here is a subjective process relying on the judgment of the user and his or her knowledge of the watershed to arrive at a "reasonably well calibrated" model. No mention has been made here of more mechanistic or "rigorous" calibration methods. Most of these methods use optimization techniques to arrive at a best fit. In the author's opinion these are impractical for the vast majority of applications and actually quite unnecessary. They are available for only a few models and it is unlikely that in a practical (and probably fairly complicated) application they could outperform the user who is well versed in hydrology, hydraulics, and water quality. These methods are really only effective when there are few parameters to manipulate and few goodness-of-fit criteria. In most urban stormwater modeling studies we are trying to fit event volumes, pollutant loads, hydrographs, and pollutographs to the true situation. Setting up a formal optimization procedure to come up with a "best fit" that accounts for all of these different needs would be difficult indeed.

A simple calibration exercise for the Oakdale Avenue model is shown in Table 4. In this case, just two independent storms were used and the focus was on matching the runoff volumes and peak flows. This is a relatively unsophisticated calibration exercise. A more complete analysis might look at the entire hydrograph and, if possible, use more than two storms. However, if the modeling objectives were only concerned with runoff volumes and peak flows, this level of effort becomes more appropriate.

**Table 4. Calibration Exercise, Oakdale Avenue Watershed**

Storm 1 — May 19, 1959 (0.68 in.)

| | | Percent of Estimated Values | | | | | | | |
|---|---|---|---|---|---|---|---|---|---|
| Run# | | 1 | 2 | 3 | 4 | 5 | 6 | 7 | 8 |
| Imperviousness | | 100 | 90 | 95 | 96 | 96 | 96 | 96 | 96 |
| Width | | 100 | 100 | 100 | 100 | 80 | 60 | 50 | 55 |
| Pipe Roughness | | 100 | 100 | 100 | 100 | 100 | 100 | 100 | 100 |
| | Actual | Simulated Values | | | | | | | |
| Volume, in. | 0.27 | 0.28 | 0.25 | 0.26 | 0.27 | 0.27 | 0.27 | 0.26 | 0.27 |
| Peak, cfs | 7.3 | 8.8 | 8.1 | 8.4 | 8.5 | 8.0 | 7.5 | 7.1 | 7.3 |

Storm 2 — July 26, 1960 (0.80 inches)

| | | Percent of Estimated Values | | | |
|---|---|---|---|---|---|
| Run# | | 9 | 10 | 11 | 12 |
| Imperviousness | | 96 | 110 | 120 | 120 |
| Width | | 55 | 55 | 55 | 80 |
| Pipe Roughness | | 100 | 100 | 100 | 100 |
| | Actual | Simulated Values | | | |
| Volume, in. | 0.45 | 0.31 | 0.36 | 0.39 | 0.39 |
| Peak, cfs | 4.3 | 2.9 | 3.3 | 3.5 | 3.7 |

Storm 1 — May 19, 1959 (0.68 inches)

| | | Percent of Estimated Values | |
|---|---|---|---|
| Run# | | 13 | 14 |
| Imperviousness | | 120 | 120 |
| Width | | 80 | 80 |
| Pipe Roughness | | 100 | 150 |
| | Actual | Simulated Values | |
| Volume in. | 0.27 | 0.33 | 0.33 |
| Peak, cfs | 7.3 | 9.6 | 8.3 |

Storm 2 — July 26, 1960 (0.80 inches)

| | | Percent of Estimated Values | |
|---|---|---|---|
| Run# | | 15 | |
| Imperviousness | | 120 | |
| Width | | 80 | |
| PIpe Roughness | | 150 | |
| | Actual | Simulated Values | |
| Volume in. | 0.45 | 0.39 | Calibrated yet? |
| Peak, cfs | 4.3 | 3.5 | |

All changes in model parameters were made uniformly throughout the watershed (e.g., if imperviousness was increased to 120% of its original value it was changed by this percentage in all 13 subwatersheds). This is actually quite reasonable since flow data were collected only at the outlet of the watershed. Making changes in one subwatershed, while not making the same change in others, would not be justified. In spite of the simplifications, what is shown here will illustrate the process well.

The process began with a storm for May 19, 1959 and the three "important" parameters set at the values estimated by Brandstetter. The estimated runoff volume was a little high (0.28 vs. 0.27 inches) and as a result imperviousness was reduced in run 2 to 90% of the starting values. This had the effect of underestimating the volume, so imperviousness was adjusted to 95% and then 96% of the original values in runs 3 and 4. This produced a good estimate. Changing imperviousness by as much as ±20% was felt to be within the bounds of possible error. However, since imperviousness is fairly easy to measure, wider variations from the original estimates seemed unjustified.

The simulated peak flow for the May 19th event was higher than its real counterpart. A change in the width of overland flow seemed appropriate. A modest reduction in run 5 to 80% of the original estimates produced a lower peak but not quite low enough (it also reduced the volume slightly). Width was lowered further in runs 6, 7, and 8 to where the simulated peak matched the measured peak. The volume remained matched as well. Note that this parameter was altered more freely since there was less confidence in the original estimates. (The width of overland flow was originally estimated to be twice the average width of the watershed; see Table 1.)

It is easy to calibrate a model to one storm; the true test comes with a second storm. A second storm, from July 26, 1960, was used to continue the exercise. Run 9 used the parameter values as they stood after calibration to the first storm (run 8). The results were not encouraging. The volume was severely underestimated, as was the peak flow. An increase in imperviousness would raise both (as was seen in the sensitivity analysis done earlier), but this may cause problems with the first storm. Nevertheless, run 10 investigated this possibility by raising imperviousness to 110% of its original value. The volume and peak were raised substantially, but not quite to the levels indicated by the actual storm data. Imperviousness was raised to 120% of the starting values in run 11, bringing both volume and peak closer to the measured values. However, a further increase in this parameter seemed unwarranted, so in run 12 the width of overland flow was brought back up to 80% of the original estimates.

With enough effort, the model could be calibrated quite well for the second storm. But what would be the effect of these parameter changes on the first storm? Run 13 shows the results for the first storm using the parameters values from run 12. Not surprisingly, the estimates were high, so pipe roughness was changed to 150% of the original values in order to bring the peak down a bit. A check on the effect on the second storm (run 15) showed that this change reduced the peak slightly.

It appears that a compromise has been reached. While the model moderately overestimates the volume and peak for the first storm, it underestimates both for the second storm. So what happens now? One could stop and say that the model performs adequately since the deviations are not "excessive" (a term that can only be defined in the eye of the beholder). On the other hand, the exercise could continue and perhaps produce a more credible model. However, is it worthwhile or even possible? (It is possible that there are errors in the rainfall and runoff data that will make significant improvements unlikely!)

The most salient point of this exercise is that the calibration of a model is an iterative process that almost always involve compromises. More time and effort may produce better results but will also increase the investment in model development. Again, the wise user will not go beyond the point of decreasing net benefits.

### 3.3.3 Problems with Precipitation Data

Precipitation data drive the urban stormwater model. Unfortunately, the precipitation record is a major source of uncertainty *and* bias in many modeling studies. More often than not, this is because too few rain gages have been established to properly capture the spatial variability of precipitation events. There is no shortage of potential problems. A single gage may be located such that orographic effects cause it to consistently report higher amounts of rainfall than are occurring over the majority of the watershed. A single gage, or a very small number of gages, can make it appear as though an intense thunderstorm that rumbled across a large watershed never occurred!

The proper design of rain gage network is a matter of debate. Practical issues only complicate the matter. Common sense goes a long way and, as always, so does a sharp eye on the modeling objectives, available resources, and the model's capabilities. A planning study might only need annual average rainfall from a reasonably close rain gage. A detailed design study may need storm hyetographs from many gages located in or very near the watershed.

An accurate picture of the spatial variability of rainfall is especially important for models used to simulate a large watershed to a high level of detail (e.g., for design purposes). The validation of such a model will be very difficult without data from a well-designed rain gage network. Attention to spatial variability is needed for validation even if the goal is to apply a hypothetical design storm to the entire watershed.

Spatial variability is simulated by allowing hyetographs to be input at various locations in the watershed. For example, SWMM does this by assigning each subwatershed to one of as many as ten different hyetographs. Some models have similar capabilities; others can only accept one hyetograph. In the latter case, such a model should probably only be used for planning purposes or for detailed studies on fairly small watersheds.

On the design of a rain gage network, the Water Pollution Control Federation (now the Water Environment Federation) Manual of Practice FD-17 (1989) offers the following:

> There are several schools of thought regarding the density of a rain gage network. In Canada, rainfall and collection system modelers recommend one gage every 1 or 2 km (0.6 or 1 mile). In Britain, the Water Research Centre has recommended only half that density, or one gage every 2 or 5 km (1 or 3 miles). In the U.S., current spacing recommendations are related to thunderstorm size. The average thunderstorm is 6 to 8 km (4 to 5 miles) in diameter and round in "footprint". Therefore rain gages are frequently spaced every 6 to 8 km on centers. Optimum rain gage spacing is still a matter of some conjecture...
>
> Two or more gages are probably needed for catchments that exceed a few hundred hectares in size, especially for convective (thunderstorm) rainfall events. Gages should be distributed so that equal areas and representative topography are sampled.

The rainfall process is continuous and smooth (even though some downpours may seem to be instantaneous). Yet, most precipitation data are presented in "quantized-continuous" form (Bedient and Huber, 1992). Typically, precipitation data take the form of average intensities over fixed time intervals (e.g., inches per hour per 5-minute interval) or depths over fixed time intervals (e.g., inches per 15-minute interval). This can present problems when the time interval is too large for the watershed being simulated. For example, an hourly rainfall record would probably not be adequate for a small urban watershed of less than a few hundred acres. A typical rainstorm will change rapidly over an hour. A reported rainfall of, for example, 0.5 inch over an hour may actually have occurred over the span of 10 minutes. The small urban watershed will be responsive to the short, intense rainfall event, but a model using the reported rainfall will never have a chance.

A brief discussion of sources for precipitation data (and other meteorological data) can be found in Chapter 4.

### 3.3.4 Problems with Quality Simulation

In a critique of the urban stormwater field, Sonnen (1980) claimed that the state-of-the-art in modeling urban stormwater quality was fairly dismal. The charge was perhaps exaggerated, but there remains a kernel of truth in what he had to say.

Many studies have successfully validated models to simulate the hydrologic and hydraulic behavior of urban stormwater. Far fewer can cite comparable success with respect to stormwater quality. Most model parameters and algorithms used to generate pollutant loads are somewhat abstract and difficult to link with measurable, physical parameters. Few would disagree that quality simulation adds considerable effort and difficulty to modeling studies.

What does this all mean? Foremost, users must be very careful when performing urban stormwater quality simulations (Huber, 1986). Most urban stormwater models are capable of producing reasonably accurate quantity simulations after a modest validation effort. This is unlikely to occur with urban runoff quality simulation. Urban runoff models must be validated very carefully and thoughtfully for quality simulation. Even then, the results may prove to be disappointing.

One question often goes begging. Is the extra effort to build a credible stormwater quality model justified? Sometimes the answer is no. For example, in the author's experience it has occasionally proven to be more effective to focus the modeling effort on stormwater quantity and to superimpose actual pollutant concentration data to predict loads. After all, the delivered pollutant load is essentially limited by its transport mechanism (i.e., urban stormwater in this case). This approach has its drawbacks (especially when trying to predict stormwater quality under future conditions), but can the picture painted be a good quantity model and some actual quality data be any worse than a questionable stormwater quality model "calibrated" to those same data?

Stormwater quality modeling is improving all the time and it is often used successfully. The above argument, while running the risk of overstatement, is designed to convince the reader, yet again, to keep the modeling effort in sight of the objectives.

## 3.4 ASSESSING UNCERTAINTY

A topic that is often not adequately addressed in modeling is the effect of uncertainty. Systematic error or bias is reduced, if not removed, in the validation process. Uncertainty, on the other hand, is difficult to remove, but the user must at least quantify it in some way so that the model results can be viewed with the proper perspective.

In practical situations the distinction between bias and uncertainty may be difficult to see. In theory, the validation process removes bias. However, are we certain it has been removed? Probably not. Because of this, the discussion that follows the term "uncertainty" should be interpreted to include some doubt about the validation results.

Sensitivity analysis is an effective tool for assessing uncertainty. This topic has already been addressed in the context of an assessment of which model parameters are most "important" in terms of their impact on model output (see Section 3.3.2). Sensitivity analysis is performing the same service here, except now we are looking at how errors in various model parameters, input variables, or initial values of the state variables (what were collectively called model factors earlier) might affect the output of the validated model.

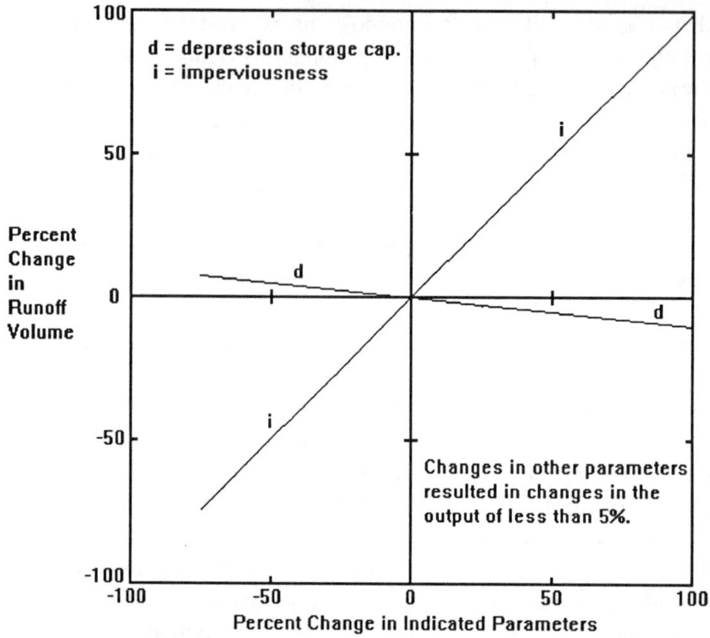


**Figure 12. Post-calibration sensitivity analysis of Oakdale Avenue catchment model, runoff volume.**

Sensitivity analysis changes one model factor while holding all others constant and observes the changes in model output. Typically, "sensitivity" is presented as the percent change in the model output vs. the percent change in the model factor. The percent variability in a particular model factor should reflect an assessment of likely error. Model output that fluctuates widely in response to this variability is sensitive to uncertainty in that model factor. This sounds rather qualitative but the user can quantify the effect by stating that a ±X% error in a model factor will result in a ±Y% change in model output.

The Oakdale Avenue application (with the storm of May 19, 1959) serves as a good example. Figure 12 shows the sensitivity of runoff volume to percent changes from the *calibrated values* for imperviousness as well as the *original values* for depression storage capacity. (Recall that the depression storage capacity was not changed during calibration because of the model's lack of sensitivity to it.) Changes in all other parameters caused insignificant changes. Figure 13 shows the sensitivity of the runoff peak to percent changes for several parameters. Not surprisingly, the results in Figures 12 and 13 are not much different from what was shown in Figures 9 and 10. The notable exception is that the changes in the peak flow shown in Figure 13 are somewhat suppressed from their counterparts in Figure 10 when pipe roughness or

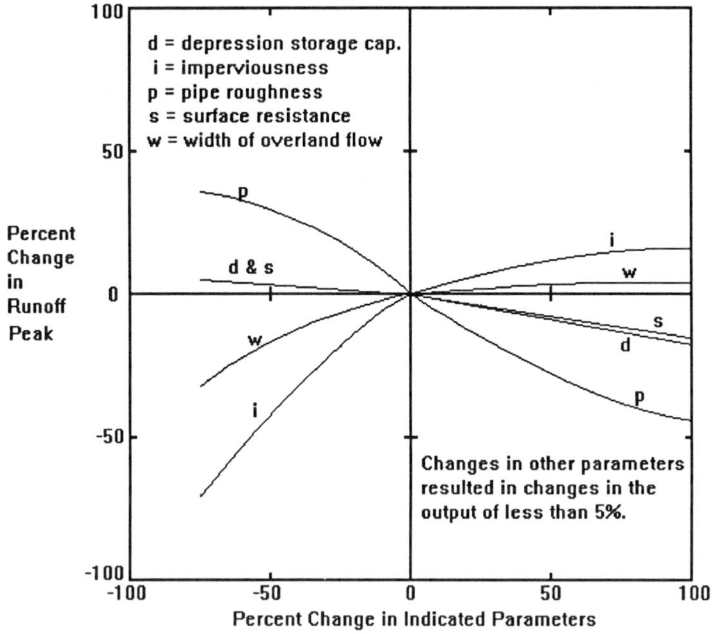


**Figure 13. Post-calibration sensitivity analysis of Oakdale Avenue catchment model, runoff volume.**

imperviousness is altered. Some of this difference is due to a small to moderate amount of surcharging, which the RUNOFF Block is not capable of modeling adequately (essentially the peaks are "clipped" in a manner similar to that shown in Figure 28 in Chapter 2). That this happened is illustrative and shows, once again, that the user must be aware of how the model works. Without this knowledge the user comes to some erroneous conclusions.

Post-calibration sensitivity analysis provides a firmer basis (as opposed to the pre-calibration effort) for analyzing the sensitivity of the model results to random error since the validation process presumably reduced the bias. However, sensitivity analysis has limitations. The effects of interactions between parameters are not addressed and the user has little information about the statistical nature of parameter error. Even with these limitations it can be very useful, particularly when a modeling study is under review. Sensitivity analysis can deflect criticism. A particular reviewer may question the validity of the model results if he or she feels a particular parameter was incorrectly specified. Sensitivity analysis may show that the "error" perceived by the reviewer has insignificant impact. Of course, it is not wise to rely only on the results of sensitivity analysis. It should go without saying that some good hydrology, hydraulics, and environmental science should be used to respond to criticisms.

## 3.5 SUMMARY

- The beginning of an urban stormwater modeling study must focus on identifying the conditions under which the study will be conducted and defining the goals and objectives of that study. Selecting the right model requires careful thought. A good rule of thumb is to select the simplest model capable of meeting the study objectives.
- All urban runoff models are imperfect representations of the urban watershed. Because of this they must be massaged into a credible tool. The processes of verification and validation are used to accomplish this.
- Two kinds of error appear in modeling studies: bias and uncertainty. A credible model must deal with both. Systematic error is reduced through validation. The effect of random error or uncertainty can be assessed with sensitivity analysis.

## REFERENCES

Bedient, P.B. and W.C. Huber, *Hydrology and Floodplain Analysis*, 2nd ed., Addison-Wesley, Reading, MA, 1992.

Brandstetter, A., *Assessment of Mathematical Models for Storm and Combined Sewer Management*, EPA-600/2-76-175a, U.S. Environmental Protection Agency, Cincinnati, August 1976.

Fleming, G., *Computer Simulation Techniques in Hydrology*, Elsevier, New York, 1975.

Huber, W.C., "Modeling Urban Runoff Quality: State-of-the-Art," in *Proceedings of Engineering Foundation Conference on Urban Runoff Quality — Impact and Quality Enhancement Technology*, Urbonas, B. and L.A. Roesner, Eds., American Society of Chemical Engineers, New York, 1986, pp. 34–48.

Huber, W.C. and R.E. Dickinson, *Storm Water Management Model, Version 4, User's Manual*, EPA-600/3-88-001a, U.S. Environmental Protection Agency, Athens, GA, 1988.

McCuen, R.H. and Snyder, W.M., *Hydrologic Modeling: Statistical Methods and Applications*, Prentice-Hall, Inc., Englewood Cliffs, NJ, 1986.

Moffa, P.E., S.J. Nix, P.L. Freedman, J.K. Marr, and R. Field, *Control and Treatment of Combined-Sewer Overflows*, Van Nostrand Reinhold, New York, 1990.

Metcalf and Eddy, Inc., University of Florida, and Water Resources Engineers, Inc., *Storm Water Management Model, Volume — Final Report*, EPA Reports 11024 DOC 07/71, U.S. Environmental Protection Agency, Washington, DC, July 1971.

Shannon, R.E., *System Simulation: The Art and the Science*, Prentice-Hall, Englewood Cliffs, NJ, 1975.

Sonnen, M.B., "Urban Runoff Quality: Information Needs," *J. Tech. Councils*, Vol. 106, No. TC1, August 1980, pp. 29–40.

Water Pollution Control Federation (Water Environment Federation), *Combined Sewer Overflow Abatement, Manual of Practice FD-17*, 1989.

# 4 USING MODELS TO MAKE PREDICTIONS

## 4.1 OVERVIEW

Ultimately, practical applications of urban stormwater models focus on making predictions for assessment, planning, or design purposes. Earlier in this book, an analogy was made with laboratory experiments. Continuing with the analogy, we can say that the validated model, that is, the experimental setup, is ready to conduct simulation experiments. The next step is the design, execution, and analysis of those simulation experiments.

## 4.2 SIMULATION EXPERIMENTS

Simulation experiments are generally divided into two groups: those run under present conditions and those run under assumed future conditions (i.e., watershed conditions such as land use, drainage system, population, and pollution abatement measures). Both groups are important. Model runs (i.e., the simulation experiments) under present conditions can be used to develop a "base line" for identifying and assessing a problem and against which to compare the results of model runs under future conditions. Model runs under future conditions are, of course, used to develop the predictions around which planning and design studies depend.

Beyond this, the experimental design has to meet the needs of the study of which it is a part. As discussed in Chapter 3, urban stormwater studies fall along a continuum of complexity that can be roughly divided into three groups:

preliminary assessment, planning, and detailed analysis and design. The following sections discuss some of the modeling issues associated with each of these.

## 4.3 PRELIMINARY ASSESSMENT STUDIES

Preliminary assessment studies are intended to place an order of magnitude on urban stormwater problems, construct a prioritization of problems, and perhaps suggest abatement strategies. In this case, the simulation experiments are quick and easy. Using the Storm Water Management Model — Level I as an example (see Chapter 2), one can see that conducting a new simulation experiment is a simple matter of changing one or more of a very limited set of inputs and parameters. The output of this model consists of average annual runoff volumes (depths, actually) and pollutant loads. This kind of information requires little in the way of sophisticated analysis. The user only really needs to remember that these simple models are rarely, if ever, validated on a site-by-site basis nor are they intended to be (typically they consist of empirical formulas developed from a broad data base). This means that the results they produce are best used for comparative analysis (From where are most of the pollutants generated? Which areas would be most suitable for abatement activities?).

## 4.4 PLANNING STUDIES

Planning studies very often involve long-term continuous simulation and models that have simple routing capabilities (the "second level" models discussed in Chapter 2). Typically, the objective is to determine the long-term impact of changes to the watershed and/or abatement strategies. The focus is sometimes on the statistical nature and frequency of various "event" characteristics such as volume, flow rate, and pollutant concentration or load. At other times the interest is in the long-term behavior of the system. Many years of hourly rainfall data are commonly used as input to the model. (The form of these data has, for the most part, been dictated by the manner in which they are published by governmental agencies.)

Three issues come to the forefront: (1) how to define an "event", (2) how long of a period to use continuous simulation, and (3) what errors arise from inadequate precipitation data. The following three sections address these issues. Later sections contain brief discussions of meteorological data sources and frequency analysis, a tool often used in planning studies.

### 4.4.1 Defining Events

An *event*, in this context, is defined as an independent episode. As with most definitions there is at least one undefined or vague term. In this case, it is the term "independent". The need for independence between events in crucial to many of the statistical analyses performed in planning studies. The focus here will be on independent runoff events. However, receiving water events could also be identified, such as episodes of oxygen depletion or elevated bacterial counts.

Identifying independent rainfall events has been an issue in the literature for many years and a number of methods have been proposed (Eagleson, 1970; Tavares, 1975; Howard, 1976; Medina, 1979; Heaney et al., 1977; Hydroscience, 1979; Restrepo-Posada and Eagleson, 1982; Nix, 1982). The transferability of these methods to the identification of independent runoff events may be problematic but no such difficulties will be assumed here.

There is perhaps no best method, but the most widely accepted approach is to divide independent events by a minimum amount of "dry" time, i.e., the minimum interevent time. Any activity not preceded by this minimum is not part of a new independent event but still part of the event containing the previous occurrence of precipitation or runoff.

Several techniques have been proposed to determine the minimum interevent time or dry period. Medina (1979), Howard (1976), and Heaney et al. (1977) recommend that an open-series autocorrelation analysis be performed on hourly rainfall data to determine the time lag at which no significant autocorrelation is present (as measured by the autocorrelation coefficient). This lag becomes the minimum dry period. Autocorrelation is essentially the correlation between data at one point in time and data occurring some time earlier (i.e., the lag time).

An autocorrelogram (a plot of the autocorrelation coefficient vs. lag time) prepared by Medina (1979; and in Heaney et al., 1977) for the 1968 hourly rainfall record in Des Moines, IA is shown in Figure 1. The 95% tolerance limit is indicated in the figure and was chosen to indicate significant autocorrelation. In this case, the 95% tolerance limit was first reached at 7 hours but Medina recommended selecting the point where the autocorrelation coefficient first reaches a minimum, which happened to be zero. This gave a minimum dry period of 10 hours.

Medina also analyzed the autocorrelation for the runoff simulated by STORM (see Chapter 2) with the same year of hourly rainfall data. The resulting autocorrelogram is shown in Figure 2. In this case, the appropriate minimum dry time separating independent runoff events was deemed to be 9 hours. The shorter time is due to the fact that STORM first satisfies depression storage and evaporation before producing runoff as a simple product of a runoff coefficient and the rainfall rate.

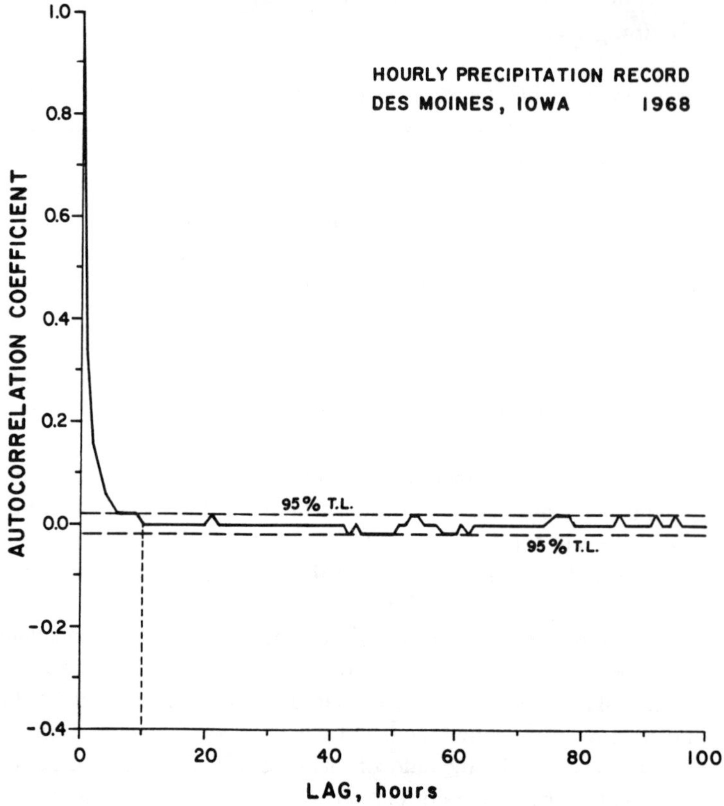


**Figure 1. Lag-k autocorrelation function of Des Moines, IA, hourly precipitation, 1968 (after Medina, 1979).**

A second approach examines the relationship between the minimum dry period and the number of events over an extended period of time (Heaney et al., 1977). Several such relationships are plotted in Figure 3. The point where an increase in the minimum dry time no longer causes a significant decrease in the number of events determines the proper value. This approach is more subjective than the autocorrelation technique; as seen in Figure 3, a well-defined breakpoint is not always evident.

Yet another approach assumes that the times between independent events, or interevent times, are exponentially distributed (Hydroscience, Inc. 1979). The exponential distribution and its discrete counterpart, the Poisson distribution, are commonly used to describe times between independent, random events (Benjamin and Cornell, 1970). The mean and standard deviation are equal for the exponential distribution and, as a result, the coefficient of variation (the standard deviation divided by the mean) is 1.0. Given this assumption, the minimum dry period applied to a long sequence of, say, hourly rainfall data

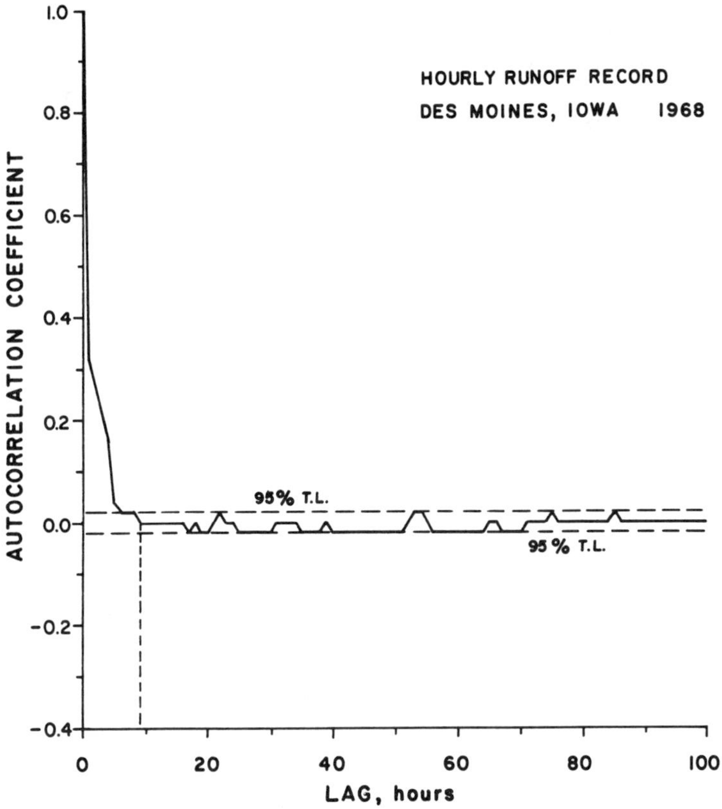


**Figure 2. Lag-k autocorrelation function of Des Moines, IA, hourly runoff, 1968 (after Medina, 1979).**

is varied until the coefficient of variation for the interevent times is 1.0. This method was used in the Nationwide Urban Runoff Program (USEPA, 1983) to determine the appropriate minimum dry period for rainfall events at a number of cities around the U.S.

### 4.4.2 Selecting an Appropriate Simulation Run Time

What is the best simulation run time for predictive purposes? The longer, the better. However, there is always some sort of limiting factor and usually it is the length of the rainfall record. Since most continuous simulation studies seek to produce information about the frequency of various events, the more appropriate question might be "How long must the simulation be to produce *useful* frequency information?" Viessman et al. (1989) claim that, in general, "frequency analysis should be avoided when working with records shorter than 10

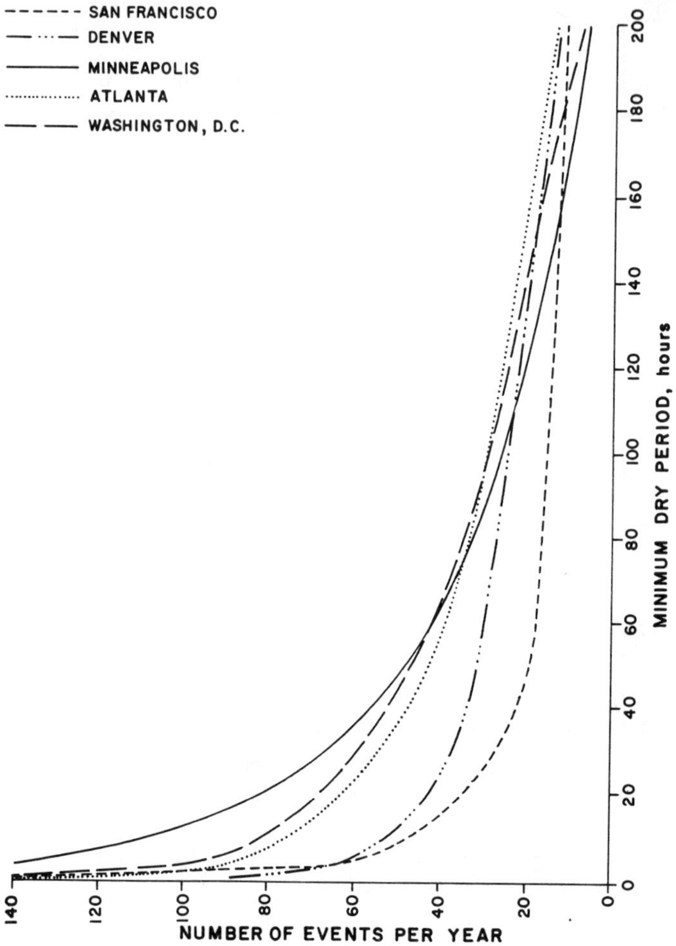


**Figure 3. Number of events per year vs. the minimum dry period for several cities (after Heaney et al., 1977)**

years and in estimating frequencies of expected hydrologic events greater than twice the record length." The "records" they were referring to are actual field data but there is no reason to think that the same rule of thumb could not apply to simulated data.

### 4.4.3 More Problems Caused by Inadequate Precipitation Data

Some of the problems associated with inadequate precipitation data were discussed in Chapter 3. However, some further discussion is warranted here.

The conditions under which a model is used to conduct continuous simulation should be the same conditions under which it was validated. Following this simple rule helps to ensure that the model remains valid during the simulation experiments. A model is not only calibrated with actual data values. It is also implicitly calibrated under the set of conditions by which those data were collected. A model calibrated using hourly precipitation at rain gage A has accounted for the location of the gage. The use of records from a different gage location or records quantized in a different manner (i.e., 15-minute values instead of hourly values) would violate the calibration.

The discussion above hopes for the ideal, but some latitude is necessary. For validation and experimentation purposes, a long, continuous precipitation record located in the watershed is good. Long, continuous records from several locations in and around the watershed is better. However, for simulation experiments used in planning studies, the principal interest is the frequency of various runoff and pollution events rather than the historical sequence of events. A compromise that is sometimes necessary is to validate the model with shorter spans of continuous precipitation data, e.g., one or two years, and to conduct the experiments with much longer records from a nearby rain gage. If the nearby gage experiences the same meteorological and climatological patterns as the gage used in calibration, this is probably superior to conducting a frequency analysis with just a few years of events simulated with a short, on-site record.

### 4.4.4 Long-Term Meteorological Data Sources

Long-term meteorological records from National Weather Service (NWS) stations around the U.S. are most readily obtained from various agencies of the National Oceanographic and Atmospheric Administration (NOAA). Daily values for precipitation, evaporation, temperature, and a number of other climatological factors, along with monthly averages and extremes, are published in *Climatological Data*. This document is published monthly and annually by the Environmental Data Service (EDS) of NOAA and can be obtained for a specific state. Hourly precipitation data, along with daily amounts, for first-order NWS stations are published in *Hourly Precipitation Data*. This is also published monthly and annually by the EDS and can be obtained for a specific state. Both of these documents are commonly available at university and municipal libraries. Hourly precipitation data are also available on computer storage media from the NOAA National Climatic Data Center, Asheville, North Carolina 28801-2696 (704-259-0682). Data for one station or a number of stations can be obtained and the cost is very reasonable. Similar data are available from the Atmospheric Environment Service in Canada. (15 minute-interval rainfall data are also becoming available.)

The sources discussed above are the prime, but certainly not the only, sources of input data. A number of governmental agencies, universities, and public and private utilities also publish meteorological data, although generally on a much less extensive basis.

### 4.4.5 Frequency Analysis of Simulation Results

The true value of continuous simulation comes from the ability to perform frequency analyses on the model results. Frequency analysis allows the user to estimate the return period or recurrence interval of various event characteristics such as volume or pollutant load.

A *frequency curve* depicts the probability that an event greater than or equal to a specified level will occur during some time period (Viessman et al., 1989). In addition, it can show the return period of an event that is equal to or greater than a specified level. The *return period* or *recurrence interval* is the average period of time between the events of interest and is calculated as

$$T = 1/P \tag{4.1}$$

where $P$ = probability that an event equal to or greater than a specified level will occur during a given time period (e.g., a year), time$^{-1}$

$T$ = return period or recurrence interval, number of time periods (e.g., number of years)

The definitions given above assume an interest in the exceedance of a specified value. They could have also been written from the perspective of being less than or equal to a specified value (note that 1–$P$ is the cumulative probability). For the remainder of this discussion the focus will be on exceedance, unless otherwise stated.

Ranking the set of events produced by a continuous simulation run from largest to smallest and keeping only the top n events, where n is the number of time periods in the total simulation time, creates an *exceedance series* (e.g., the largest 20 events in 20 years.) From this series, $P$ for each event can be estimated as

$$P = m/(n+1) \tag{4.2}$$

where $m$ = rank of the event of interest (with $m$ = 1 being the largest)

$n$ = the number of events in the exceedance series

Using Equation 4.1, the return period can then be estimated by

$$T = (n+1)/m \tag{4.3}$$

This estimate of the return period is known as the *plotting position*. Equation 4.3 is actually just one of many formulas for the plotting position, but this one, known as the Weibull formula, is probably the most commonly used (Viessman

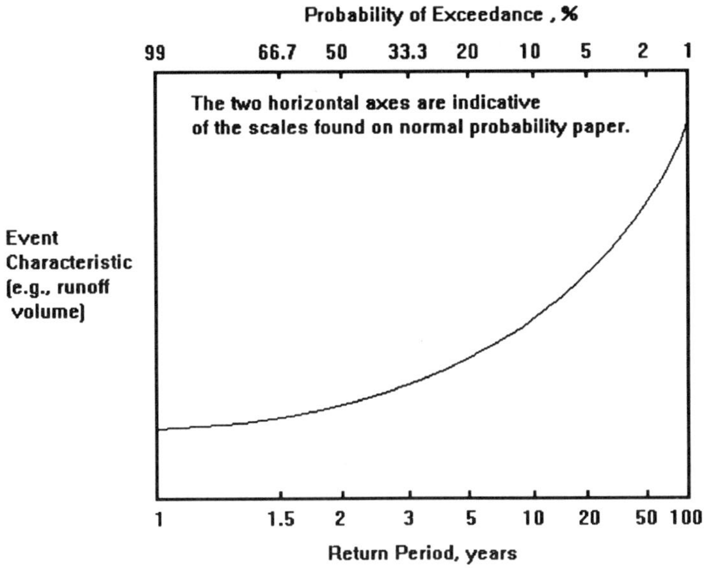


**Figure 4. Frequency curve.**

et al., 1989). Plotting the estimates of $T$, $P$, or $(1–P)$ for each event in the exceedance series against the size of the event creates a frequency curve. An example is shown in Figure 4.

The development of a frequency curve can be facilitated by the use of probability paper. There are several kinds of probability paper. All have one axis for plotting the event values and a second for probability. Normal probability paper, the most common and often referred to as simply probability paper, has an arithmetic axis to plot the event value and a probability scale arranged such that if the events are normally distributed the plot will appear as a straight line. This is a convenient feature, especially when the object is to determine if a normal cumulative distribution is appropriate for the data. However, probability paper can simply be used as a convenient plotting medium (regardless of the distribution of the data). Other forms of probability paper are designed for other cumulative distributions (e.g., log-normal probability paper, which has a logarithmic scale for the event values and results in straight line plots for data that fit the log-normal distribution).

### 4.4.6 Example Applications

Figures 5 and 6 show examples of frequency analysis applied to urban runoff quantity and quality. These curves were developed from a long, continuous simulation of the runoff and total suspended solids loads generated by a hypothetical 640-acre watershed. The RUNOFF Block of SWMM, driven by

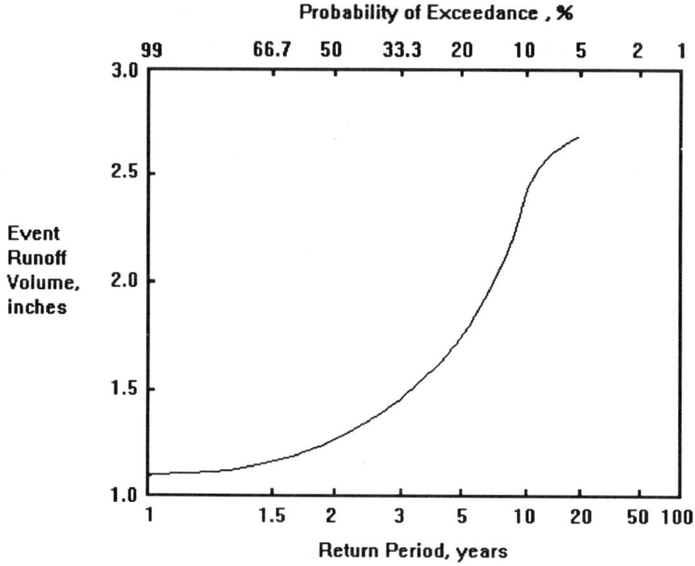


Figure 5. Frequency curve, event volumes, hypothetical watershed.

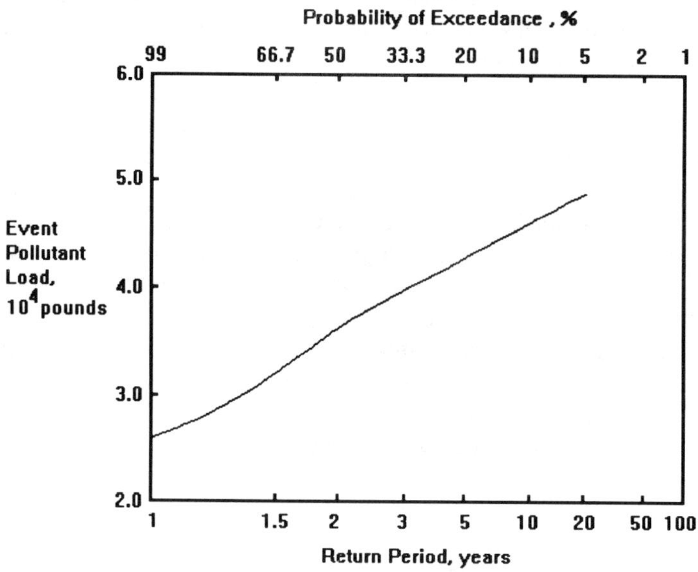


Figure 6. Frequency curve, event suspended solids loads, hypothetical watershed.

**Table 1. Characteristics of Hypothetical Watershed, SWMM RUNOFF Block**

| | |
|---|---|
| Area | 640 acres |
| Slope | 0.01 ft/ft or 1% |
| Imperviousness | 50% |
| Width of overland flow | 2000 ft |
| Depression storage capacity | |
| Impervious areas | 0.08 in. |
| Pervious areas | 0.20 in. |
| Surface resistance, Manning's n | |
| Impervious areas | 0.013 |
| Pervious areas | 0.350 |
| Horton infiltration parameters | |
| Minimum infiltration rate | 0.45 in./hr |
| Maximum infiltration rate | 2.50 in./hr |
| Infiltration decay rate | 0.00115/s |

20 years of hourly precipitation for Syracuse, NY, was used to simulate the flows and loads. (The author is fully aware that much of the precipitation in Syracuse does not fall as rain, but for the purposes of illustration, it was assumed to occur this way.) The characteristics of the watershed are shown in Table 1. With a minimum dry period of 6 hours the STATISTICS Block (see Chapter 2) identified 1326 independent *runoff* events. This period was selected because it produced a coefficient of variation of 1.0 for the interevent times (see discussion concerning the identification of independent events in Section 4.4.1). The curves shown in Figures 5 and 6 depict the exceedance probabilities and return periods for the event runoff volumes and the event total suspended solids loads, respectively. This kind of information can be helpful in characterizing the nature of the runoff and pollutant loads generated by the urban area. In addition, control strategies can be developed by targeting events equal to or smaller than a critical return period (e.g., providing enough retention storage to contain the 2-year event) or by producing event volume and pollutant load frequency curves for different control strategies.

Another approach, based on the long-term behavior of the watershed, is also possible. A study by Nix and Heaney (1988) provides a useful example. In this case, the STORAGE/TREATMENT Block of SWMM was set up to simulate pollutant removal in a detention basin receiving the runoff from a hypothetical 640-acre watershed (much like the one described in Table 1). The SWMM RUNOFF Block was to used to create a year-long sequence of runoff and pollutant loads from the watershed. (The 1969 hourly rainfall record for Atlanta was used as input. This was considered to be, statistically, a "typical year" for Atlanta and, thus, used to avoid long simulation runs.) The detention basin was assumed to have a constant release rate and to operate such that flows

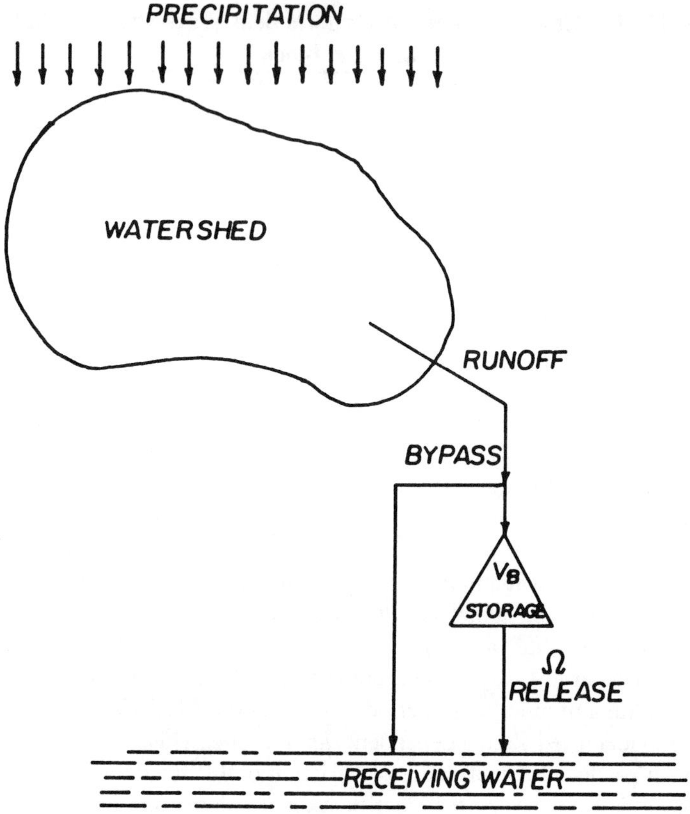


**Figure 7. Stormwater storage-release system (from Nix and Heaney, 1988).**

bypass the basin when it is full. The general configuration is shown in Figure 7. The basin was also assumed to behave as a completely mixed reactor and to have vertical sides, a depth of 10 feet, and a rectangular surface area with a length-to-width ratio of 2. An unspecified pollutant was given first-order decay characteristics and a rate constant of 1.0 per day (this being deemed as fairly representative of many pollutants and pollution indicators).

The STORAGE/TREATMENT Block was run for 36 combinations of storage capacity (varied through the surface area) and release rate. The resulting long-term pollutant removal percentages are shown in Figure 8. Isoquants of pollutant removal are also shown. The units for the storage capacity and the release rate are normalized over the watershed area (i.e., millimeters and millimeters per hour, respectively). This information, along with cost data, can be used to develop optimal strategies for different pollutant removal levels using the techniques discussed in Section 2.6.3.

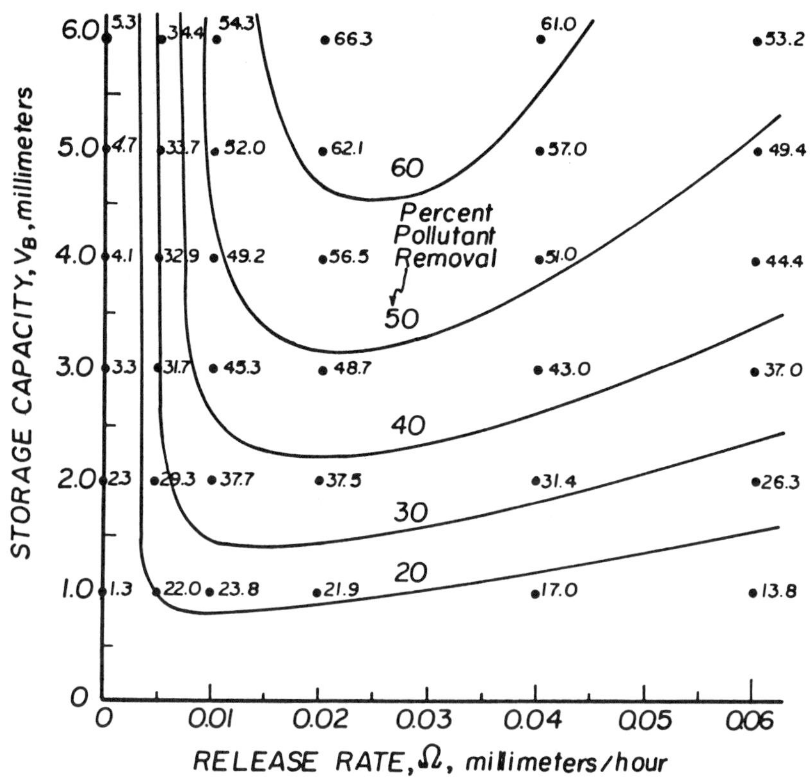


**Figure 8. Long-term storage-release system performance (from Nix and Heaney, 1988).**

### 4.4.7 Linkage to the Receiving Water

The analysis of receiving water quality is an important part of many stormwater studies. This text does not deal with receiving water modeling, but it is important for the reader to understand how the coupling of an urban runoff model and a receiving water model can address the impact of urban runoff and possible abatement strategies on a receiving water. This will be accomplished with an example application developed by Heaney et al. (1977) and Medina (1979).

STORM was set up to simulate a year-long series of runoff events from Des Moines, IA. Hourly rainfall data for 1968 were used to drive the model. Using the autocorrelative method discussed in Section 4.4.1, 9 hours was selected as an appropriate minimum number of dry hours to separate events. The Des Moines metropolitan area is served by combined sewers and separate sewers

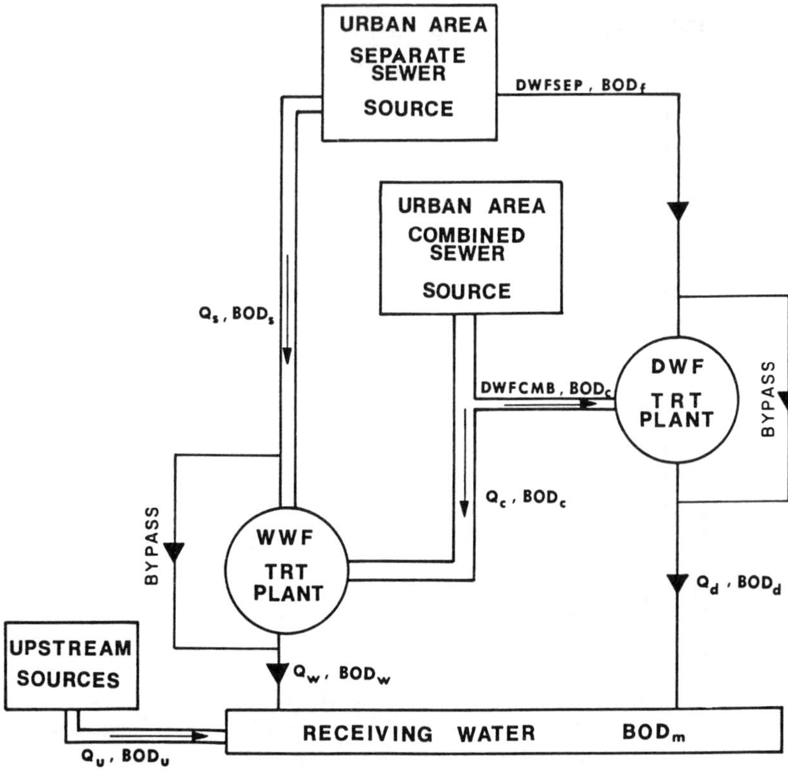


**Figure 9. Comprehensive urban stormwater control configuration (Medina, 1979).**

and thus several simple assumptions were used to divide the STORM-generated flows and BOD loads into the contributions from separate and combined areas. Dry-weather flows and loads were estimated with data from dry-weather point sources in the area.

Wastewater discharges from Des Moines find their way into the Des Moines River and its tributary, the Raccoon River (and various other small contributing streams). A simple model to simulate the effect of dry- and wet-weather wastes on dissolved oxygen levels in receiving waters was constructed and applied to the Des Moines River for the conditions occurring during 1968. The BOD loads produced with STORM were linked to the receiving water model to investigate several abatement alternatives.

When this analysis was performed only sanitary flows and intercepted combined flows in Des Moines were treated at the secondary level (i.e., about 85% BOD removal); other wet-weather flows were released untreated. The more comprehensive system shown in Figure 9 was used to evaluate other

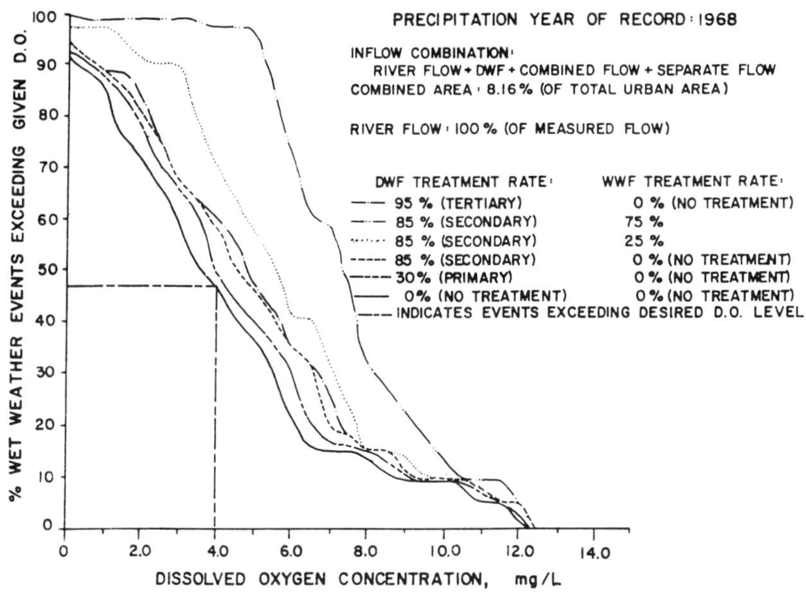


**Figure 10. Minimum DO frequency curves for varied treatment alternatives, Des Moines, IA (from Medina, 1979).**

strategies by assuming fixed percentages of BOD removal at hypothetical dry-weather and wet-weather treatment plants. Figure 10 shows the percentage of wet-weather events causing the exceedance of various dissolved oxygen levels. (This is just another form of frequency analysis.) The results show that improving treatment at the dry-weather plant to the tertiary level (i.e., 95% BOD removal) has little effect on river water quality during wet weather. The combination of a wet-weather treatment facility (even at a fairly low level) along with conventional secondary dry-weather treatment results in far fewer violations of the dissolved oxygen target level of 4.0 mg/L.

## 4.5 DETAILED ANALYSIS AND DESIGN STUDIES

Detailed analysis and design studies often involve models in which the system is depicted in considerable spatial and temporal detail. Typically, models are used to developed "critical" hydrographs (and perhaps pollutographs) with which to analyze and/or design conveyance and abatement systems. In some cases, only the peak flow is needed (e.g., for sewer design) and in others the entire hydrograph is required (e.g., for routing through storage).

### 4.5.1 The Design Event

In most design studies, one or more "design" rainstorms are used to drive a model for the purpose of generating the critical runoff hydrographs. The problems associated with design rainstorms used in this manner were discussed briefly in Chapter 3 and other discussions can be found in the literature (McPherson, 1978; Adams and Howard, 1985; Nix, 1982; Voorhees and Wenzel, 1982; Niemczynowicz, 1984; Huber and Dickinson, 1988). Again, the main problem is that the frequency characteristics of an input (i.e., precipitation event) to a complex system such as a watershed will not be reproduced by the output (i.e., runoff or pollution event). Regardless, this is a venerable method that will remain popular for some time, if for no other reason than the proliferation of quantity and quality control regulations that are based on design rainstorms.

The selection and development of an appropriate design event is an interesting topic and discussions can be found in several basic textbooks (Chow et al., 1989; Viessman et al., 1989; Bedient and Huber, 1992). It is useful, however, for the reader to remember that the fundamental principle in selecting a design storm is that it should maximize net benefits. This concept is illustrated in Figure 11. For many, if not most, studies, the design event is preordained by tradition or regulation. These design storms are often too conservative and frequently result in expensive, wasteful designs. They should always be challenged.

### 4.5.2 Example Application

Heaney et al. (1975) describe an application of SWMM to analyze design changes in the Stevens Avenue combined sewer system located in Lancaster, PA. The model indicated that surcharging was occurring in the system during the storm of interest. Several conduits were enlarged until the surcharging no longer occurred. The hydrographs at the Stevens Avenue overflow are shown in Figure 12 for the existing system and the altered system. As the figure shows, the suppression of surcharging comes at the price of an increase in the peak overflow discharge; something that has, of course, other implications.

### 4.5.3 The Role of Continuous Simulation

Why is continuous simulation not used in this kind of work? It would seem to make more sense to analyze the hydrologic, hydraulic, and pollutional

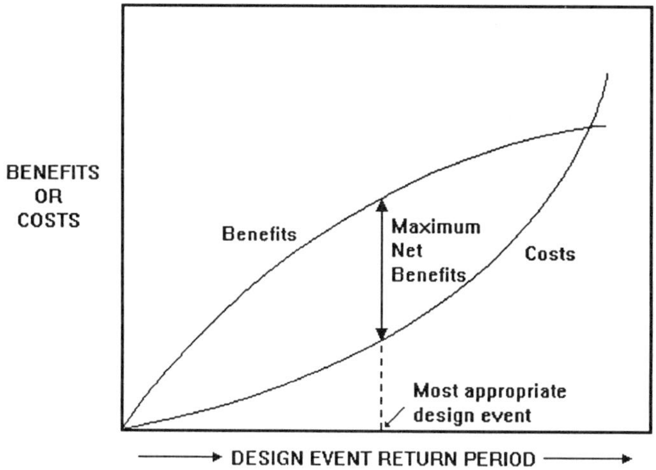


**Figure 11. Design event maximizing net benefits.**

behavior of different designs over the course of, for example, 50 years. There are at least two reasons. First, computers still require considerable time to do lengthy simulations on detailed systems (several hours of computer time is possible). Second, long-term precipitation data are not conveniently available in intervals short enough for design work. When analyzing the hydraulics of a drainage system, for example, only short-interval precipitation data (i.e., 5-minute intervals) will be able to produce hydrographs that accurately predict peak flows. As computers, modeling techniques, data bases, and data base technology advance, models that depict the urban watershed in great detail will become more and more practical and continuous simulation will very likely be used for design.

A possible compromise involves the use of continuous simulation to identify critical *historical* runoff events or periods which can then be used in a detailed modeling study for design purposes. This approach is described in several publications (Robinson and James, 1984; Huber et al., 1986; Bedient and Huber, 1992). One of the advantages of this method is that the design is based on something from experience, that is, the designer can point to the fact that his or her design will control the notorious "storm of July 20, 1992" — probably a considerable comfort to the public. The disadvantage, of course, is that the long historical record is just one of an infinite number of possible records that could have occurred over the same period and the "storm of July 20, 1992" might be a truly important event or just one of the larger storms from a rather humble sequence.

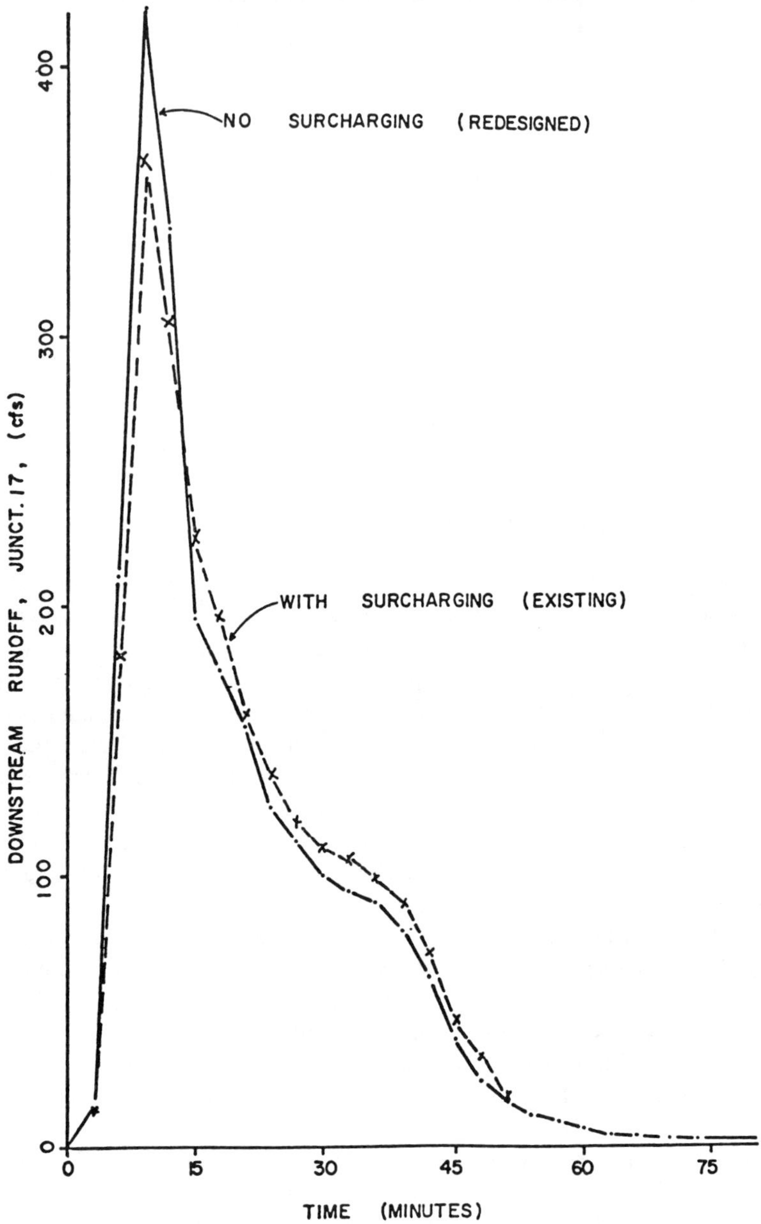


**Figure 12.** Effect of surcharging and conduit alterations on Stevens Avenue overflow (from Heaney et al., 1975).

## 4.6 ACCOUNTING FOR UNCERTAINTY

In addition to the uncertainty associated with input errors and model structure errors discussed in Chapter 3, the prediction phase of a modeling study must be concerned with the uncertainty associated with the future. Mechanically, the same sensitivity analysis tools discussed in Chapter 3 can be used to quantify the uncertainty associated with future land use patterns through the model parameters (e.g., changing imperviousness to reflect a change from single-family residential to multi-family residential). The question then becomes "Does the uncertainty associated with input and model errors overwhelm the uncertainty associated with the future?" If this is so what can one really say about the predictions generated by the model? Probably not much in an absolute sense, but in a relative sense probably still quite a bit.

## 4.7 SUMMARY

- Experimentation with a computer model requires good technique and solid analysis, just at it does in a laboratory.
- Urban runoff modeling studies can generally be classified into three groups: preliminary assessment, planning, and detailed analysis and design.
- Preliminary assessment studies use models that require little input data and little in the way of sophisticated calculation. They produce long-term average runoff volumes and pollutant loads (typically average annual volumes and loads).
- Planning studies focus on the frequency characteristics of urban runoff events or the long-term behavior of the watershed. Frequency analysis is used to determine the return periods of event characteristics developed from long, continuous simulations.
- Detailed design studies generally focus on developing critical hydrographs at one or more points in the urban watershed for the purposes of designing conveyance and abatement facilities. The "critical" hydrographs are selected to represent particular return periods.
- All modeling studies are concerned with predicting the effects of different control strategies or a changing watershed. When dealing with the latter, an additional sort of uncertainty is added to the picture — the future of the watershed. Essentially, the same sensitivity analysis techniques used to assess input error can be used to assess different futures.

## REFERENCES

Adams, B.J. and C.D.D. Howard, *The Pathology of Design Storms*, Publication 85-03, University of Toronto, Department of Civil Engineering, Toronto, Ontario, January 1985.

Bedient, P.B. and W.C. Huber, *Hydrology and Floodplain Analysis*, 2nd Ed., Addison-Wesley, Reading, MA, 1992.

Benjamin, J.R. and C.A. Cornell, *Probability, Statistics and Decision for Engineers*, McGraw-Hill, New York, 1970

Chow, V.T., D.R. Maidment, and L.W. Mays, *Applied Hydrology*, McGraw-Hill, New York, 1989.

Eagleson, P.S., *Dynamic Hydrology*, McGraw-Hill, New York, 1970.

Heaney, J.P., W.C. Huber, H. Sheikh, M.A. Medina, J.R. Doyle, W.A. Peltz, and J.E. Darling, *Urban Stormwater Management Modeling and Decision Making*, EPA-670/2-75-022, U.S. Environmental Protection Agency, Cincinnati, May, 1975.

Heaney, J.P., W.C. Huber, M.A. Medina, Jr., M.P. Murphy, S.J. Nix, and S.M. Hasan, *Nationwide Assessment of Combined Sewer Overflows and Urban Stormwater Discharges: Volume II, Cost Assessment*, EPA-600/2-77-064, U.S. Environmental Protection Agency, Cincinnati, March 1977.

Howard, C.D.D., "Theory of Storage and Treatment-Plant Overflows," *J. Environ. Eng. Div.,* Vol. 102, No. EE4, August 1976, pp. 709–722.

Huber, W.C., B.A. Cunningham, and K.A. Cavender, "Use of Continuous SWMM for Selection of Historic Design Events in Tallahassee," *Proc. Stormwater and Water Quality Model Users Group Meeting*, Orlando, EPA-600/9-86-023, U.S. Environmental Protection Agency, Athens, GA, March 1986, pp. 295–321.

Huber, W.C. and R.E. Dickinson, *Storm Water Management Model, Version 4, User's Manual*, EPA-600/3-88-001a, U.S. Environmental Protection Agency, Athens, GA, 1988.

Hydroscience, Inc., *A Statistical Method for the Assessment of Urban Stormwater*, EPA-440/3-79-023, U.S. Environmental Protection Agency, Washington, DC, May 1979.

McPherson, M.B., *Urban Runoff Control Planning*, EPA-600/9-78-035, U.S. Environmental Protection Agency, Washington, DC, October 1978.

Medina, M.A., Jr., *Level III: Receiving Water Quality Modeling for Urban Stormwater Management*, EPA-600/2-79-100, U.S. Environmental Protection Agency, Cincinnati, August 1979.

Niemczynowicz, J., "Can the Rainfall Input Be Modified So That Frequencies of Rainfall-Input and Runoff-Output Will Be Similar?," *Water Sci. Technol.*, Vol. 16, 1984, pp. 251–254.

Nix, S.J., Analysis of Storage/Release Systems in Urban Stormwater Quality Management, Ph.D. dissertation, University of Florida, Gainesville, 1982.

Nix, S.J. and J.P. Heaney, "Optimizaiton of Stormwater Storage-Release Strategies," *Water Resour. Res.*, Vol. 24, No. 11, November 1988, pp. 1831–1838.

Restrepo-Posada, P.J. and P.S. Eagleson, "Identification of Independent Rainstorms," *J. Hydrol.*, Vol. 55, 1982, pp. 303–319.

Robinson, M.A. and W. James, Chedoke Creek Flood Storage Computed by Continuous SWMM and Dynamic Storms, Report No. R124, Computational Hydraulics, Inc., Department of Civil Engineering, McMaster University, Hamilton, Ontario, December 1984.

Tavares, L.V., "Continuous Hydrological Time Series Discretization," *J. Hydraul. Div.*, Vol. 101, No. HY1, January 1975, pp. 49–63.

U.S. Environmental Protection Agency, *Results of the Nationwide Urban Runoff Program, Volume I, Final Report*, NTIS PB84-185552, U.S. Environmental Protection Agency, Washington, DC, December 1983.

Viessman, Jr., W., Lewis, G.L., and Knapp, J.W., *Introduction to Hydrology*, 3rd ed., Harper and Row, New York, 1989.

Voorhees, M.L. and H.G. Wenzel, "Sensitivity and Reliability of Design Storm Frequency," *Proc. Second Conference on Urban Storm Drainage*, Urbana, IL, 1982, pp. 374–383.

# 5 URBAN RUNOFF MODELING: WHERE DOES IT GO FROM HERE?

## 5.1 OVERVIEW

There is little doubt that computer models have been indispensable to the solution of many urban stormwater problems. There also is little doubt that much of dubious quality has been produced with these same models and that there is room for improvement. This chapter begins by probing some of the weaknesses of models, especially the large models that most think of when the topic of urban runoff modeling arises. The chapter concludes with some suggestions for future directions in the field.

## 5.2 THE SEVEN SINS OF RUNOFF MODELS

A paper by Douglass B. Lee, appearing in the 1973 volume of the *Journal of the American Institute of Planners*, had the ominous title "Requiem for Large-Scale Models". Professor Lee was examining "fundamental flaws in attempts to construct and use large, comprehensive urban land-use models," but his discourse was and is applicable to large, comprehensive urban runoff models (Nix, 1990). He also presented a lengthy discussion as to why these models (the land-use planning models) suffered their demise not long after they were born, although he seemed to suggest that their downfall was temporary and simply part of a cycle. This sort of cycling has occurred to some degree in urban runoff modeling as the field has gone through several small booms and retrenchments (sometimes because the models moved ahead of the data base available to support them). While the early optimism (circa 1970s) about the ability of computer models to help solve urban runoff problems has generally ebbed, they remain important tools.

Lee pleaded that the lessons learned in the development and use of large models not be lost. This is good advice. At the core of Lee's critique are seven "sins" allegedly committed by large urban planning models, including hypercomprehensiveness, grossness, hungriness, wrongheadedness, complicatedness, mechanicalness, and expensiveness. The discussion that follows uses Lee's seven sins, with some poetic license, to develop some useful insights and suggest new directions. Two models in particular, the Storm Water Management Model (SWMM) and the Hydrological Simulation Program-Fortran (HSPF), come under some criticism. This should not be viewed as an attack on these software packages. It is just a matter of familiarity on the part of the author.

### 5.2.1 Hypercomprehensiveness

It is useful to differentiate comprehensive scope from comprehensive detail. The former implies that a model is capable of simulating (however well or poorly) most facets of the urban runoff picture. The latter implies that a model is more sophisticated in terms of the scientific detail applied to individual runoff processes. Both kinds can be useful. Both can be problems. It is rare that a model contains both.

Model developers often attempt to create comprehensive programs capable of handling a large range of applications. Unfortunately, while these kinds of models are often comprehensive in scope, they probably will not do any one thing very well. (The analogy to a one-person band is appropriate.) This is not necessarily bad, as long as the user knows what he or she is getting. Models like these can give useful information, especially when the user is looking at the *scope* of the problem.

Models purporting to be comprehensive in detail (and perhaps comprehensive in scope, too) contain minutiae that are often unsubstantiated or not very useful. Detail gives the appearance of sophistication but often provides little additional information. Again, this kind of comprehensiveness is not necessarily bad, but the user should understand and accept the modeling concepts. In addition, users *and* developers should question the cost-effectiveness of the sophistication.

SWMM is the probably the quintessential large urban runoff model (Huber and Dickinson, 1988; Roesner et al., 1988). As we saw in Chapter 2, SWMM contains several tens of thousands of lines of code and is designed to simulate the movement of runoff and pollutants from the ground surface, through pipe and channel networks and storage/treatment facilities, and to receiving waters. SWMM intends to be comprehensive in scope and, to a degree, comprehensive in detail. However, some of that detail is marginally useful. Two instances are particularly illuminating. The first is an algorithm designed to estimate the quantity and quality of sanitary sewage from urban areas (meant to be useful in modeling combined sewer systems). The routine is based on a large amount of demographic data provided for a number of subareas that are not coincidental with the subwatersheds (see Chapter 2). These flows are added to the surface runoff and routed through the sewer system. Much of the information the algorithm requires is difficult to obtain and the effort goes into estimating flows that may be insignificant during wet weather. Reasonable estimates of sanitary sewage flow and pollutant loads can probably be estimated much more easily.

A second instance of questionable detail is the algorithm used to simulate surface runoff pollutant loads. A large portion of the model documentation is devoted to describing interesting, but not well-proven, methods for simulating the accumulation and washoff of surface pollutants. The detail is impressive and the programming is extensive. On the other hand, the detail may not be worth the cost. The results are likely to be less accurate than pollutant loads generated by applying concentrations obtained by a minimal sampling program to the flows generated by the model (Moffa et al., 1990). This is not to say that this algorithm is useless; it can work well with careful application and an abundance of calibration data (Huber, 1986). The argument here is that the level of apparent detail in the program leads the user to believe that the results are equally impressive. This is simply not the case.

### 5.2.2 Grossness

"Grossness" is a sin that is directly related to the hypercomprehensive scope of most large runoff models. In order to handle the scope *and* maintain tractability, the sophistication of the model often suffers. Therefore, most large models cannot and should not be applied to very specific tasks. In fairness though, this sin is committed most often by users applying a model to something for which it was not intended.

HSPF is another example of a large, ostensibly comprehensive model (Johanson et al., 1984). It, like SWMM, contains tens of thousands of lines of code and hundreds of subroutines (although the reason for the latter is an admirable attempt to modularize the program). HSPF is the latest in a long line of descendants from the Stanford Watershed Model, one of the original large watershed models. HSPF, unlike SWMM, is also intended for rural applications.

HSPF is designed to predict runoff quantity and quality. There is considerable detail with the runoff process compartmentalized into a maze of storages (upper zone storage, lower zone storage, ground water storage, etc.) linked by one or more flow processes (percolation, interflow, ground water flow, etc.). However, the algorithms are somewhat abstract in that, while they intuitively satisfy us, they are difficult to relate to direct physical measurements. (Just what is "nominal upper zone soil storage" or the "interflow inflow parameter"?) This level of detail again leads one to feel that the results produced are equally impressive. Whereas there are plenty of algorithms to handle various facets of the runoff process, few, if any, are capable of studying a particular facet with scientific rigor. For example, no serious student of the evapotranspiration process would use the corresponding algorithm in HSPF to study evapotranspiration.

### 5.2.3 Hungriness

Large runoff models require large amounts of data. These data are sometimes difficult to find and manage (although easy for the developer to request). The wise user will look at these data requirements in light of what the model is being asked to do. SWMM and HSPF are good examples of models that *can* demand large amounts of data to satisfy their detailed depictions of the runoff process. When using these or any other models, the user should do a quick, qualitative benefit/cost analysis. In the case of SWMM, for example, is it worthwhile to obtain demographic data to estimate sanitary sewage flow or would a short talk with a municipal engineer and some elementary calculations give you adequate estimates? Is it necessary to segment the watershed into 50 subcatchments? Would five be adequate? This has been a common theme throughout this book. Balance the modeling effort with the needs of the study.

### 5.2.4 Wrongheadedness

Each large, comprehensive runoff model is structured in such a way as to preclude certain applications and each has errors. It is probably fair to say that many model revisions are initiated by users that uncover something the model could not handle correctly. Limitations and errors are expected because any model is constrained by the biases, knowledge, and experience of its developer(s). Unfortunately, due to the size and complexity of some models, many limitations and errors go undetected.

Models such as HSPF and SWMM depend on feedback from users. Typically, errors and unintended limitations reported to the developers are quickly corrected. However, many are yet to be detected. If you are doing something new with one of these models, beware! You may be the first to test a particular algorithm adequately.

### 5.2.5 Complicatedness

Lee commented "whatever the components of a city or a model (of a) city are, their microscopic behavior is largely unknown. The best information we have has to do with aggregate relationships that include the effects of an unknown but large number of other variables." Is runoff modeling really any different? Many large models contain excruciating detail on certain aspects of the hydrologic cycle — detail that may be totally unnecessary, if not wrong. This has been discussed earlier.

However, all those details lead to many interactions among numerous model components, variables, parameters, and algorithms. The number of these relationships can be overwhelming and lead to problems. When an error is corrected, the interconnected nature of the model may lead to a ripple effect, where the correction causes seen or unseen problems elsewhere. More importantly, the developer loses touch with the model as it becomes more complex, detailed, and comprehensive. This may be the most insidious of the seven sins.

Interestingly, the kind of sensitivity analysis described in Chapter 3 often reveals that changes in one or more parameters have little or no effect on the model output. The obvious question is "Why are they there in the first place?" The obvious is sometimes never asked.

### 5.2.6 Mechanicalness

Mechanicalness refers to the fact that large runoff models are programs executed on digital computers and, as a result, there are round-off errors. It also refers to errors inherent in the use of various numerical procedures for the solution of algebraic, differential, and integral equations.

"Mechanicalness" may be the most irritating of abuses. There is no doubt that it takes considerable effort to make a computer model work properly. Constructing a large program takes considerable time and resources. All too often, though, this effort is expended in the details of making the computer do our bidding rather than on the problem itself. As Lee states, this effort "is then mistaken for intellectual activity and creative problem solving."

Most large models have groups responsible for their maintenance. In addition, there are many hundreds of engineers and scientists using and modifying them. Through all this activity, most models have not fundamentally changed, in terms of basic modeling concepts, over the course of their lives. Much of the more recent activity in model "development" has been in getting them to work on personal computers, improving the numerical procedures, or improving their "user friendliness". These efforts, which may or may not be useful, divert much needed energy from enhancing our understanding of urban runoff and of better ways to model it.

### 5.2.7 Expensiveness

Millions of dollars have been spent to develop and maintain large runoff models. Using them also requires large commitments of time and resources. Gathering the data, figuring out how to make the model work, understanding how the model works, validating the model, and conducting the simulation experiments involve considerable person-hours and computer resources.

The previous paragraph contains no surprises. But it conjures up a theme present throughout this book. Match the model to the task as much as possible. Avoid using a large model when a simpler, smaller one will suffice.

## 5.3 WHAT IS (OR SHOULD BE) AHEAD FOR URBAN RUNOFF MODELING?

The well-known runoff models are, for the most part, useful and well regarded. And rightly so. The developers of these models have provided us with a very good set of tools. The conscientious user can manipulate them to do some credible analysis and design work. However, most large models suffer from one or more of Lee's seven sins. They are often described as too complicated, too expensive, and not of the "real world". The developers and maintainers of these models can do some things to better serve the user community.

Developers should learn what the users are doing with their models and, if necessary, redesign them. For example, the majority of SWMM users are invoking only a small fraction of the model's capability. Most use SWMM to estimate and route flows for sewer system design and combined sewer overflow abatement. Far less use is made of the quality routines or the storage/treatment routines, among other portions. Developers should create "low-overhead" models that are easily managed by the typical user and directed toward common applications. (Some enterprising vendors do this already.) Models will become more useful if they do a few needed functions well as opposed to many things less well.

Considerably more effort should be directed toward improving our understanding of the fundamental hydrologic, hydraulic, and water quality processes rather than cosmetic model features and marginally useful detail. Much effort has been put into spinning software around some fairly straightforward hydrology and hydraulics. Our *intellectual* activity should focus more on the basic understanding of the urban runoff process and less on the computer. (The reader should not mistake these comments as some sort of backlash against the computer. It is only a plea for more balance and a reordering of priorities.)

The temptation to make models too "user friendly" should be avoided. Model developers should strive for better "user–model interaction". A good model should complement the user's knowledge and judgment, not form it. Too often "user friendly" means taking the user out of the modeling process.

A good model should provide a set of tools that allow the thoughtful user *confidently* to simulate an urban watershed. The confidence comes from having the user understand the theoretical foundations of the model and involving him or her in as much of the application as possible. This does not mean that the task should not be made easier for the user. However, a poor model is one that does not ask the user to apply some judgment and to share his or her heuristic knowledge of the system being modeled. It is important for users to remember that the responsibility for the application of a model to a problem is theirs and theirs alone. The model developer can help ensure that this responsibility remains where it belongs by requiring significant user investment. (Judging by the proliferation of disclaimers at the head of most user's manuals, this is a desire shared by model developers.) The "black box" approach is unacceptable and unethical. The user must understand the scientific, engineering, and mathematical concepts employed by the model.

Perhaps the most important, yet often overlooked, part of modeling is the efficient manipulation of data and information. Huber et al. (1991), in an excellent assessment of the state-of-the-art in computer technology as applied to urban runoff, call for a more enlightened use of computers in solving urban runoff problems. At the core of their more significant recommendations is the need to better manage and use data and knowledge bases. They are right. For too many years, modeling studies have focused on the model and not the data needed to support them. Users should make the gathering, filtering, storage, retrieval, and intelligent application of watershed, runoff quantity and quality, and control technology performance data a high priority. Modern data handling technologies such as data base management systems, spreadsheets, and geographic information systems (GIS), are capable of providing the necessary "infrastructure" to manage and use data intelligently. The actual collection of local data will remain a costly proposition, but the effective use of data should, in most cases, become rather routine and inexpensive. Model developers can help the cause by providing interfaces with popular data base management software and spreadsheets. More importantly, it is probably time for serious reassessment of the way in which modeling is done. Rather than seeing the problem primarily from a conceptual point of view, it may be time to use the computer to take a more informational perspective.

## 5.4 SUMMARY

- Urban runoff models have proven to be very useful tools. However, as with any tool, an inappropriate application can produce undesirable results.
- Douglas Lee, in a very cogent article (considered a "classic" treatise by the author), identified seven sins committed by large land-use planning models. Many of these "sins" are also committed by large runoff models.

- Model developers should learn what users need and create well-focused models that do a small set of things well rather than large smorgasbord models that do many things in a mediocre fashion.
- The technical community should balance its efforts so that our understanding of the physical processes and ability to mathematically model them improves at a rate comparable to our computing technology.
- User-model interaction should be improved so that users are confident in taking full responsibility for model output.
- More focus should be placed on the information needed to study an urban runoff problems and improving ways to manage that information.

## REFERENCES

Lee, D.B., "Requiem for Large-Scale Models," *J. Am. Inst. Planners*, Vol. 39, No. 3, March 1973, pp. 163–178.

Huber, W.C., "Modeling Urban Runoff Quality: State-of-the-Art," in *Proceedings of Engineering Foundation Conference on Urban Runoff Quality — Impact and Quality Enhancement Technology*, Urbonas, B. and L.A. Roesner, Eds., American Society of Chemical Engineers, New York, 1986, pp. 34–48.

Huber, W.C. and R.E. Dickinson, *Storm Water Management Model, Version 4, User's Manual*, EPA-600/3-88-001a, U.S. Environmental Protection Agency, Athens, GA, 1988.

Huber, W.C., J.P. Heaney, T.G. Curtis, and T.L. Leathers, *Review of State-of-the-Art of Integration of Modern Microcomputer Technology with Models for Management of Storm and Combined Sewer Runoff Quality*, Purchase Order No. VN1-310-420000, U.S. Environmental Protection Agency, Edison, NJ, 1991 (draft).

Johanson, R.C., J.C. Imhoff, J.L. Kittle, and A.S. Donigian, *Hydrological Simulation Program — FORTRAN (HSPF): Users Manual for Release 8.0*, EPA-600/9-80-015, U.S. Environmental Protection Agency, Athens, GA, 1984.

Moffa, P.E., S.J. Nix, P.L. Freedman, J. Marr, and R. Field, *Control and Treatment of Combined-Sewer Overflows*, Van Nostrand Reinhold, New York, 1990.

Nix, S.J., "'Requiem for Large-Scale Models' Revisited," Proceedings of Symposium on Transferring Models to Users, Janes, E.B. and W.R. Hotchkiss, Eds., American Water Resources Association, Bethesda, MD, November 1990, pp. 321–328.

Roesner, L.A., J.A. Aldrich, and R.E. Dickinson, *Storm Water Management Model, Version 4, User's Manual: EXTRAN Addendum*, EPA-600/3-88-001b, U.S. Environmental Protection Agency, Athens, GA, 1988.

# ANNOTATED INPUT DATA GUIDE FOR THE EPA STORM WATER MANAGEMENT MODEL

## A.1 GENERAL COMMENTS

The EPA Storm Water Management Model (SWMM) was discussed extensively in Chapter 2. This appendix provides additional, practical information to help with the application of the most used portions of the popular operational blocks: RUNOFF, TRANSPORT, and EXTRAN.

SWMM is, without doubt, the preeminent urban stormwater management model. However, SWMM is over 20 years old and it is beginning to show its age and the effects of a rather piecemeal maintenance program. Essentially, SWMM is a FORTRAN program designed in the mainframe environment of the late 1960s and early 1970s. It has undergone three major revisions since being introduced in 1971 (Metcalf and Eddy, Inc. et al., 1971) and has been altered enough to make it workable on personal computers. The latest "sub-release" available at the time of this writing (Version 4.2) is accompanied with the documentation issued with Version 4.0 in 1988 (Huber et al., 1988; Roesner et al., 1988) and "on-disk" files that update the input data format and inadequately describe the many new patches added since then. The latest version has a very frayed feel to it. This is not the fault of the developers or even the maintainers — they are fighting a battle with very limited resources. Through it all SWMM remains the model of choice for many in the stormwater management field.

Many of the "patches" referred to above are attempts to add marginal capabilities to SWMM. In other words, they add little to what the large majority of users are trying to accomplish. Actually, this has long been a problem with SWMM—the "sin of hypercomprehensiveness" (see Chapter 5).

This appendix has two basic objectives. The first is to peel away many of the marginal capabilities and attributes that can make SWMM seem so onerous. The second is to provide some insights for users attempting to piece together the input data files. What is "important" in the model will vary from user to user. However, a core group of capabilities important to the majority of users can probably be identified.

The following assumptions and limitations will govern the rest of this guide:

*RUNOFF Block*

- Only surface runoff quantity and quality will be simulated (i.e., ground water contributions will not be simulated).
- Only simple rainfall input (constant interval) will be applied to the model. No snowmelt will be simulated.
- Drainage/sewer systems will not be modeled with the RUNOFF Block (the TRANSPORT or EXTRAN Block will be used for this).

*TRANSPORT Block*

- Only sewer conduits with the regular shapes provided by the program will be modeled (i.e., no natural/irregular channels or user-supplied channels will be used).
- All flows and pollutant loads entering the drainage/sewer system will be generated by the RUNOFF Block. In other words, the option to allow direct input of hydrographs and pollutographs by the user will not be used.
- The program routines that generate sanitary sewage flows will not be used. The user enters constant flows and loads at various points in the system, if desired.
- The deposition and scour of solids from sewer bottoms will not be simulated.

*EXTRAN Block*

- Only sewer conduits with the regular shapes provided by the program will be modeled (i.e., no natural/irregular channels will be used).
- All flows and pollutant loads entering the drainage/sewer system will be generated by the RUNOFF Block. The option to allow direct input of hydrographs and pollutographs by the user will not be used.

The input data requirements for the "lean and mean" Storm Water Management Model are described in Sections A.2, A.3, and A.4. The reader should note that changes are constantly being made to SWMM by the model developers and a growing group of consultants and software developers. Thus, some of what appears here will be outdated almost immediately. In addition, many users are working with older versions that do not have some of the capabilities mentioned in this guide. Most users, however, will find the older and newer versions remarkably similar for the majority of applications and find this guide of some benefit. *This guide is meant to accompany the SWMM documentation, not replace it. The user is strongly urged to obtain all appropriate manuals and to review the large body of literature related to SWMM.* The SWMM user's manuals are: Huber, W.C. and R.E. Dickinson, *Storm Water Management Model, Version 4, User's Manual,* EPA-600/3-88-001a, U.S. Environmental Protection Agency, Athens, GA, 1988; and Roesner, L.A., J.A. Aldrich, and R.E. Dickinson, *Storm Water Management Model, User's Manual, Version 4: Addendum I, EXTRAN,* EPA-600/3-88-001b, U.S. Environmental Protection Agency, Cincinnati, August 1988.

### A.1.1 The Role of the EXECUTIVE Block

The EXECUTIVE Block is the traffic director of SWMM. Its most important role is to transfer data from one block to another, e.g., transferring surface flows from RUNOFF to the drainage system modeled by EXTRAN. How to do this depends on the SWMM release in hand and whether the model is implemented on a mainframe or a personal computer. Thus, the details of file transfer (interfacing) are not addressed here. Refer to the main SWMM User's Manual as well as the documentation that comes with any particular release for the specifics.

### A.1.2 General Input Guidelines

The SWMM User's Manual and its Addendum describes some general guidelines for data entry. Follow them closely. A few others are in order:

- In more recent versions of SWMM, data may be entered in "free format". In other words, a parameter does not have to be placed in particular set of columns, as in older versions. Nevertheless, the user will find that keeping parameters lined up in columns, especially when a line is repeated several times, will help locate input errors.
- Some parameters on a line may require entries even if they are not needed by the program. Zeroes or empty sets of quotes are often used to do this. It is not always clear when this is necessary. Beware! However, it is usually safe to err on the side of entering values for all parameters on a line. In addition, some input lines are omitted entirely under certain circumstances. In this guide the word "omit" means that the line should NOT be listed in the input.

### A.1.3 Default/Ratio Option

An interesting and useful, but very confusing, feature in SWMM is the "default/ratio" option. In several instances, a repetition of the same input line is used to enter data for groups of subcatchments, conduits, junctions, etc. Each of these lines has a parameter to assign an "ID" number. With the default/ratio option, this parameter can be given the values –1 or –2, instead of the normal ID number, to indicate that what follows on that line is either a ratio or default value, respectively, to be applied to those data entries on all subsequent lines. The input parameters to which this option can be applied are indicated by asterisks in the input data lines, e.g., DIST*.

With the ratio alternative (–1), the user can apply a multiplier to each of these parameters. For example, if the user wanted to increase every value for conduit length by 50% (a strange choice, but it's only an example), a ratio of 1.5 should be specified in the spot reserved for this parameter. Let us look at an example from the TRANSPORT Block. Below is part of line E1 repeated several times, each describing a different sewer conduit (DIST is the length of the conduit):

| | NOE | NUE(1) | NUE(2 | NUE(3) | NTYPE | DIST* | GEOM1* | ..... |
|---|---|---|---|---|---|---|---|---|
| E1 | –1 | 0 | 0 | 0 | 0 | 1.5 | 1.2 | ..... |
| E1 | 1003 | 1005 | 0 | 0 | 1 | 500 | 4.0 | ..... |
| E1 | 1007 | 1009 | 0 | 0 | 1 | 300 | 3.0 | ..... |

The values for DIST will be multiplied by 1.5, the ratio indicated in the ratio line. GEOM1 will be multiplied by 1.2, and so on. This will continue for all subsequent E1 lines until another ratio line is encountered. This feature can be very useful in sensitivity analysis and model calibration. (A ratio of 0.0 will cause the program to assume a multiplier of 1.0.)

With the default alternative (–2), the user can change the default values of the eligible parameters (i.e., those with asterisks) from what the program assumes to what the user desires. (Default values are called up when a zero is entered for the parameter value.) For example, if the user wanted to specify a default value of 0.013 for Manning's roughness coefficient, this value would be entered on the default line in the spot reserved for this parameter. Let us look at another example from the TRANSPORT Block. Below is yet another part of line E1 repeated several times; each time describing a different sewer conduit (ROUGH is Manning's roughness coefficient):

| | NOE | NUE(1) | NUE(2) | NUE(3) | NTYPE | ..... | ROUGH* | ..... |
|---|---|---|---|---|---|---|---|---|
| E1 | –1 | 0 | 0 | 0 | 0 | ..... | 0.013 | ..... |
| E1 | 1003 | 1005 | 0 | 0 | 1 | ..... | 0.0 | ..... |
| E1 | 1007 | 1009 | 0 | 0 | 1 | ..... | 0.0 | ..... |

The values of ROUGH for conduits 1003 and 1007 will be 0.013, the default value indicated in the default line. Of course, the default value can always be superseded by a non-zero value. The default will remain for all subsequent E1 lines until another default line is encountered. This feature can be a very useful tool in model calibration and avoiding the repetition of identical numbers for a large number of entries.

## A.2 RUNOFF BLOCK INPUT DATA

NOTE: This input data guide was modified from the material provided with SWMM Release 4.2. Underlined values are those needed to meet the assumptions listed on page 148 or otherwise suggested to simplify SWMM. Some material found in the model documentation is omitted if it is not germane to this "pared-down" version on SWMM. General comments about a line are found under the heading for that line. Specific comments follow the individual data items. General comments about a group of lines appear before that group.
In several cases the user is given the option of identifying a subcatchment or sewer system element with a name (in single quotes) instead of a number. This option may not work. Beware!

*A1 GROUP : TITLE LINES*

TITLE : Two lines with heading to be printed on output.

*****************************************************
* There must be two lines, each in single quotes ('Run No. 1'). *
*****************************************************

*B1 LINE : CONTROL DATA*

*****************************************************************
* This line sets some of the run control parameters and RUNOFF Block options, as do lines B2 and B3. *
*****************************************************************

METRIC : Metric input-output.
= 0, use U.S. customary units.
= 1, use metric units. Metric input is indicated in brackets [ ] in remainder of this guide.

ISNOW : Snowmelt parameter.
= <u>0</u>, snowmelt not simulated.
= 1, single event snowmelt simulation.
= 2, continuous snowmelt simulation.

```
****************************************************
* The snowmelt routines in RUNOFF are rarely used. The     *
* snowmelt data lines are omitted from the remainder of this *
* guide, i.e., ISNOW = 0.                                   *
****************************************************
```

NRGAG : Number of hyetographs (rain gages). Up to 10 hyetographs may be used.

```
****************************************************
* This parameter allows the introduction of as many as 10   *
* rain gages. The use of multiple gages will greatly         *
* enhance modeling success. The rain data are entered with  *
* the E3 lines.                                             *
****************************************************
```

INFILM : Choice of infiltration equation.
= 0, Horton equation used.
= 1, Mein-Larson version of the Green-Ampt equation used.

```
****************************************************
* Horton's equation was originally developed to depict the   *
* reduction in infiltration capacity in the presence of ample *
* surface moisture. An integrated form of the equation has   *
* been used here to account for the reduced loss of capacity *
* when a light rain occurs. The Mein-Larson version of       *
* the Green-Ampt equation is a two-stage model that more     *
* directly accounts for actual infiltration. Either method   *
* requires considerable attention and the user should first  *
* determine the importance of infiltration to the hydrologic *
* system being simulated.                                    *
****************************************************
```

KWALTY : Quality (or erosion) simulated?
= 0, no.
= 1, yes.

```
****************************************************
* The quality routines in the RUNOFF Block are overly        *
* cumbersome and require considerable effort to set up and   *
* validate. In some cases, their use is probably not justified. *
* It may be more effective to use the quantity routines, which *
* are fairly reliable, along with runoff quality measurements *
* to predict pollutant loads.                                *
****************************************************
```

IVAP : Evaporation parameter.
= 0, evaporation data not entered, default rate of 0.1 inches/day [3 millimeters/day] is used.
= 1, read mean monthly evaporation data in Group F1 in units of inches/day [millimeters/day].

= 2, read mean monthly evaporation data in Group F1 in units of inches/month [millimeters/month].
= 3, read monthly evaporation data in Groups F1 and F2, in units of inches/month [millimeters/month].
= 4, read evaporation from NSCRAT (3) file as, created by the TEMP Block of SWMM.

```
****************************************************
* In nearly all cases, it is probably adequate to use the     *
* default rate of 0.1 in./day (i.e., IVAP = 0), average daily  *
* rates for each month (i.e., IVAP = 1), or average monthly   *
* totals (i.e., IVAP = 2). The remainder of this guide assumes *
* that IVAP = 0, 1, or 2.                                      *
****************************************************
```

NHR : Hour of day of start of storm (24 hour clock, midnight = 0.0).

NMN : Minute of hour of start of storm.

NDAY : Day of month of start of simulation.

MONTH : Month of start of simulation.

IYRSTR : Year of start of simulation (last two digits only, e.g., 1993 = 93).

---

*B2 LINE : MORE CONTROL DATA*

---

IPRN(1) : Print control for SWMM input.
= 0, print all input data.
= 1, do not print channel/pipe, snowmelt, subcatchment, or quality data, only control information is printed.

```
****************************************************
* It is a good idea to print all input data. The quantity of *
* information is not voluminous and input errors are         *
* easier to detect.                                          *
****************************************************
```

IPRN(2) : Print control for RUNOFF Block graphs.
= 0, plot all graphs.
= 1, do not plot hyetographs (for each gage), or inlet hydrograph (which is the sum of all the hydrographs to the inlets).

```
****************************************************
* Table 4-29 in the SWMM User's Manual presents a nice    *
* summary of the output options.                          *
****************************************************
```

IPRN(3) : Print control for output of SWMM. "Totals" below refer to precipitation, runoff, and all quality parameters. Output is printed for each

inlet selected on the M1 and M3 lines. Daily, monthly, and yearly printouts only occur if the simulation is long enough.
= 0, do not print daily, monthly, or yearly totals.
= 1, monthly and annual totals only, one year per page.
= 2, daily, monthly, and annual totals, two months per page. Daily totals are printed whenever there is non-zero precipitation and/or runoff.

```
*****************************************************
* This is useful information for long, continuous runs. The    *
* quantity of information is not excessively voluminous.       *
*                                                              *
* Table 4-29 in the SWMM User's Manual presents a nice         *
* summary of the output options.                               *
*****************************************************
```

---

*B3 LINE : MORE CONTROL DATA*

```
**********************************************************************
*  The following three parameters specify the time steps to be used under       *
*  different situations. It is difficult to suggest appropriate values. The user *
*  is reminded that the time step must be short enough to keep up with           *
*  changing conditions. A general rule of thumb is to use a time step            *
*  that is some fraction of the precipitation interval during wet periods. The   *
*  transition time step can be several times longer than the wet time step since *
*  this is used in the period shortly after precipitation has stopped. While     *
*  several things are still going on during this period, the changes are fairly  *
*  gradual and a short time step is not needed. In dry periods, well after the   *
*  runoff has subsided, the time step can be set to as long as a day or even     *
*  several days.                                                                  *
*                                                                                 *
*  For single-event simulations (or even some simulations covering several events)*
*  the transitional and dry-period time steps are not important. But the use     *
*  of the transitional and dry-period time steps will shorten long, continuous   *
*  simulations.                                                                   *
**********************************************************************
```

WET : Wet-period time step, seconds. WET must be ≥ 1 second.

WETDRY : Transition-period (between wet and dry periods) time step, seconds. WETDRY should be ≥ WET and ≤ DRY.

DRY : Dry-period time step, seconds. DRY must be ≥ WET.

LUNIT : Units of LONG (simulation length; see below).
= 0, seconds.
= 1, minutes.
= 2, hours.
= 3, days.

= 4, ending date, year/month/day, (e.g., July 30, 1987 = 870730). In this case, the program stops when this date is reached.

LONG : Simulation length (units from LUNIT). Not required if LUNIT = 4.

---

*B4 LINE : SUBCATCHMENT DEFAULT PARAMETERS*

****************************************************************************
* The B4 line is used only when the user wants to modify one of the subcatchment default parameters, otherwise omit. These are nearly always left at the default values. *
****************************************************************************

PCTZER : Percent of impervious area with zero detention (immediate runoff). The default value is 25%.

REGEN : For continuous SWMM, infiltration capacity is regenerated using a Horton-type exponential rate constant equal to REGEN*DECAY, where DECAY is the Horton rate constant read in for each subcatchment in line H1. Not required for Green-Ampt infiltration. The default value is 0.01.

---

############################################################################
# Lines C1, C2, C3, C4, and C5 are used to enter general data concerning snowmelt. They are omitted because the snowmelt option was not selected in line B1, i.e., ISNOW = 0. #
############################################################################

*C1 LINE : GENERAL SNOW INPUT DATA*

*C2 LINE : MONTHLY WIND SPEEDS*

*C3 LINE : AREAL DEPLETION CURVE FOR IMPERVIOUS AREAS*

*C4 LINE : AREAL DEPLETION CURVE FOR PERVIOUS AREAS*

*C5 LINE : AIR TEMPERATURES*

*D1 LINE : PRECIPITATION INPUT DATA OPTIONS*

ROPT : Precipitation input option.

= 0, read NRGAG hyetographs on E1, E2 and E3 lines. Rain data can be saved permanently on NSCRAT(1) using the @ function in the EXEC Block (see the SWMM User's Manual).

= 1, read processed precipitation file on the JIN file (see the

EXEC Block). This file is either from the RAIN Block (earlier saved JOUT file) or from a previous run of the RUNOFF Block (earlier saved NSCRAT(1) file). Unless blocks are run as a part of a single overall SWMM run, access to earlier saved files is through the @ function described in the EXEC Block (see the SWMM User's Manual).

```
**************************************************
*  ROPT sets the precipitation input option. In general *
*  ROPT = 0 when entering rain for a single event or    *
*  for a fairly short sequence of events. In either case *
*  each dry interval must be indicated with a zero entry. *
*  For longer simulations, the rainfall is usually obtained *
*  as a file on floppy disk from, say, the National     *
*  Climatic Center and must be "processed" before use.  *
**************************************************
```

---

*E1 LINE : PRECIPITATION CONTROL DATA*

```
************************************************************************
*  OMIT THIS LINE IF ROPT = 1 IN LINE D1.                              *
*                                                                      *
*  Entering rainfall into with the RUNOFF Block has become overly complicated *
*  in the most recent version of SWMM, at least for the vast majority of applications. *
*  Most users simply want to enter the hyetograph as a sequence of intensities *
*  at regular intervals. This is assumed here.                         *
************************************************************************
```

KTYPE : Type of precipitation input. Precipitation is in units of inches/hour [millimeters/hour] for THISTO time units (see below). Use KTIME to select units of time (see below).
= 0, read KINC precipitation values per line.
= 1, read KINC time and precipitation pairs per line.
= 2, read time and NRGAG precipitation values per line.

```
**************************************************
*  Set KTYPE = 0 to enter a sequence of constant-interval  *
*  rainfall intensities.                                   *
**************************************************
```

KINC : Number of precipitation or time/precipitation pairs per line. Enter any number if KTYPE = 2.

```
**************************************************
*  Use this to set the number of intensities per per line. For *
*  example, if hourly values are to be entered, 12 values (or  *
*  half a day) per line is logical.                             *
**************************************************
```

KPRINT : Print control for precipitation input.
= 0, print all precipitation input.

= 1, suppress all but summary of precipitation input.

*****************************************************
* It is good idea to print all rainfall data, especially when *
* processing the rainfall for the first time. *
*****************************************************

KTHIS : Variable THISTO option.
= 0, rainfall interval (THISTO) is constant.
= K, where K is the number of variable rainfall intervals entered on the E2 lines. Precipitation values outside the time frame of any variable rainfall interval uses THISTO as the rainfall interval.

*****************************************************
* Set KTHIS = 0 to enter a sequence of constant-interval *
* rainfall intensities. *
*****************************************************

KTIME : Precipitation time units.
= 0, time in minutes.
= 1, time in hours.

KPREP : Precipitation unit type.
= 0, intensity, inches/hour [millimeters/hour].
= 1, total precipitation volume over the interval, inches [millimeters].

NHISTO : Number of data points for each hyetograph.

*****************************************************
* If there are multiple hyetographs NHISTO should be long *
* enough to accommodate the longest record. The shorter records*
* must be filled in with zeros until NHISTO intervals are entered.*
*****************************************************

THISTO : Time interval between values, units of KTIME.

TZRAIN : Initial time of day of rainfall input, units of KTIME.

*****************************************************
* Set TZRAIN = 0. Use zero values for rain if the storm *
* begins later than indicated on line B1. *
*****************************************************

---

*E2 LINE : VARIABLE INTERVAL PRECIPITATION INFORMATION*

---

*****************************************************************
* Line E2 is omitted because the variable interval rainfall was not selected in *
* line E1, i.e., KTYPE = 0. *
*****************************************************************

*E3 LINES : PRECIPITATION DATA*

```
**********************************************************************
*  OMIT THESE LINES IF ROPT = 1 IN LINE D1.                          *
*                                                                    *
*  For hyetograph 1 place KINC constant-interval precipitation values on each *
*  line (recall that KTYPE = 0 and KTHIS = 0 in line E1), repeating until NHISTO *
*  values are entered. Repeat for additional hyetographs, up to NRGAG times. *
*  Begin at the time indicated by NHR in line B1. See note for TZRAIN in line *
*  E1.                                                               *
**********************************************************************
```

RAIN(1) : Rainfall intensity for the first interval, inches/hour [millimeters/hour].

. .

. .

. .

RAIN(KINC) : Rainfall intensity for the last interval on line, inches/hour [millimeters/hour].

*F1 LINE : EVAPORATION DATA*

```
**********************************************************************
*  OMIT THIS LINE IF IVAP = 0 IN LINE D1.                            *
*                                                                    *
*  Average daily evaporation values are entered for each month if IVAP = 1 *
*  (line D1) or average month totals if IVAP = 2.                    *
**********************************************************************
```

VAP(1) : Evaporation rate for month 1 (January), inches/day [millimeters/day] or inches/month [millimeters/month].

. .

. .

. .

VAP(12) : Evaporation rate for month 12 (December), inches/day [millimeters/day] or inches/month [millimeters/month].

*F2 LINES : TIME SERIES OF MONTHLY EVAPORATION VALUES*

```
**********************************************************************
*  Line F2 is omitted because the simpler evaporation options assumed in line *
*  B1 (i.e., IVAP = 0, 1, or 2) do not require it.                   *
**********************************************************************
```

```
##########################################################################
#   The runoff block has a limited capability to simulate the transport of flow    #
#   and pollutants through a simple drainage system. This capability is rarely used. #
#   The TRANSPORT and EXTRAN Blocks are much better suited to this                  #
#   task. Thus, the G1 and G2 lines, which are used to characterize drainage         #
#   system elements, are omitted from this guide.                                    #
#                                                                                    #
#   There are no "control" switches for these routines. When these lines are         #
#   omitted, flow through a drainage system is not simulated.                        #
##########################################################################
```

*G1 LINES : CHANNEL/PIPE DATA*

*G2 LINE : CONTROL STRUCTURE DESCRIPTION*

*H1 LINES : SUBCATCHMENT DATA FOR QUANTITY SIMULATION*

```
**************************************************************************
*   This line is repeated for each subcatchment. There is a limit on the number   *
*   of subcatchments (which varies from one SWMM release to another), but it can be *
*   increased (see the SWMM User's Manual).                                          *
**************************************************************************
```

JK : Hyetograph number (based on the order in which they are entered in the E3 lines).

```
*****************************************************
* Here is where the rain gage most appropriate for each    *
* subcatchment is selected.                                *
*****************************************************
```

NAMEW : Subcatchment number or name.

```
*****************************************************
* This assignment is essentially arbitrary. The program keeps *
* track of the subcatchment by assigning its own number       *
* based on the order of appearance. If a name is used it must *
* be in single quotes (e.g., 'WATERSHED1').                    *
*                                                              *
* This is a "default/ratio" indicator. The parameters below    *
* with asterisks are those to which this option may be applied.*
* See the discussion under "A.1 General Comments" at the       *
* beginning of this Appendix.                                  *
*****************************************************
```

NGTO : Channel/pipe or inlet (manhole) number of drainage.

```
*****************************************************
* Since this guide is assumes that no channels or pipes are    *
* modeled with the RUNOFF Block, this number identifies the *
* inlet to which the runoff is drained. This inlet becomes the  *
* entry  point or node in the TRANSPORT or EXTRAN        *
* Block for routing through a drainage system. As many        *
* as five subcatchments may drain to a single inlet. This can  *
* be increased with a trick involving the G1 lines but it is     *
* seldom done and ignored here.                                  *
*****************************************************
```

WW(1)* : Width of subcatchment, feet [meters].

```
*****************************************************
* This term is more properly known as the width of overland *
* flow (it is not the width of the subcatchment). As          *
* discussed in Chapter 2, is difficult to estimate. Do not    *
* waste time trying to develop or follow a complicated method*
* to calculate a value. Consistently follow a simple method   *
* and plan to adjust the width during calibration.              *
*****************************************************
```

WAREA* : Area of subcatchment, acres [hectares].

WW(3)* : Percent imperviousness of subcatchment (percent hydraulically effective impervious area).

```
*****************************************************
* This has more impact on runoff than any other parameter.   *
* Estimate it carefully and the quantity simulation will probably*
* be fairly accurate. Judiciously vary it in the calibration process.*
*****************************************************
```

WSLOPE* : Ground slope, feet/foot [meters/meter].

```
*****************************************************
* Slope does not have a large impact on runoff within a         *
* reasonable range around the true value. Estimate the          *
* slope with good care and consistency, then leave it alone.   *
*****************************************************
```

WW(5)* : Impervious area Manning's roughness.

```
*****************************************************
* Manning's n for surface roughness does not have a large    *
* impact on runoff within a reasonable range around            *
* the true value. While it should be estimated with care and   *
* consistency, it can have some effect on peak flows and      *
* play a minor role in calibration.                               *
*****************************************************
```

WW(6)* : Pervious area Manning's roughness.

```
**************************************************
* See the comment for WW(5).                     *
**************************************************
```

WSTORE1* : Impervious area depression storage, inches [millimeters].

```
**************************************************
* Depression storage does not have a large impact on runoff *
* within a reasonable range around the true value. While   *
* it should be estimated with care and consistency, it can *
* have some effect on early flows and overall runoff volume *
* and thus can play a minor role in calibration.           *
**************************************************
```

WSTORE2* : Pervious area depression storage, in. [millimeters].

```
**************************************************
* See the comment for WSTORE1.                   *
**************************************************
```

```
*****************************************************************
* The following three parameters are the last ones on line H1 if Horton's *
* infiltration equation is being used, i.e. INFILM = 0 in line B1.        *
*                                                                         *
* The importance of infiltration varies from case to case. In general, it is more *
* important for highly pervious subcatchments than highly impervious      *
* subcatchments. Use sensitivity analysis to determine the importance attached *
* to the following parameters.                                            *
*****************************************************************
```

WLMAX* : Maximum initial infiltration rate, inches/hour [millimeters/hour].

WLMIN* : Minimum (asymptotic) infiltration rate, inches/hour [millimeters/hour].

DECAY* : Decay rate of infiltration in Horton's equation, 1/seconds.

```
*****************************************************************
* The following three parameters are the last ones on line H1 if the      *
* Green-Ampt infiltration equation is being used, i.e., INFILM = 0 in line B1. *
* See the comment for the Horton equation parameters above.               *
*****************************************************************
```

SUCT* : Average capillary suction, inches [millimeters] of water.

HYDCON* : Saturated hydraulic conductivity of soil, inches/hour [millimeters/hour].

SMDMAX* : Initial moisture deficit for soil, volume air/volume voids (expressed as a fraction).

```
##########################################################################
#   The H2, H3, and H4 lines are used to simulated ground water flow. The   #
#   ground water routines in the RUNOFF Block are seldom used. These        #
#   algorithms are ignored in this guide.                                   #
#                                                                           #
#   There are no "control" switches for these routines. If these lines are omitted, #
#   ground water is not simulated.                                          #
##########################################################################
```

*H2 LINES : GROUND WATER SUBCATCHMENT DATA*

*H3 LINES : GROUND WATER FLOW COEFFICIENTS AND EXPONENTS*

*H4 LINE : MORE GROUND WATER PARAMETERS*

```
##########################################################################
#   Lines I1 and I2 are used to enter subcatchment-specific data concerning #
#   snowmelt. They are omitted because the snowmelt option was not selected in #
#   line  B1, i.e., ISNOW = 0.                                              #
##########################################################################
```

*I1 LINE : SUBCATCHMENT SNOW DATA IF ISNOW = 1*

*I2 LINE : SUBCATCHMENT SNOW DATA IF ISNOW = 2*

```
##########################################################################
#   IF WATER QUALITY IS NOT SIMULATED, I.E., KWALTY = 0 IN LINE #
#   B1, OMIT LINE J1 AND SEVERAL OTHERS THAT FOLLOW AND         #
#   SKIP TO LINE M1.                                            #
#                                                               #
#   The entry of quality data is very cumbersome and confusing. Essentially, the #
#   J1 line identifies how many quality constituents and land use types will be #
#   modeled. The J2 line establishes "dust and dirt" buildup characteristics. #
#   With the J3 card the user establishes the constituent buildup relationships (which #
#   may or may not be a function of "dust and dirt" buildup). The J4 line allows #
#   a fraction of one constituent to be added to another — this is seldom used #
#   and will not be included here.                              #
#                                                               #
#   Erosion may also be simulated in SWMM. This is a rarely used and will not be #
#   included in this guide.                                     #
#                                                               #
#   More than any other portion of the input data these lines, i.e., lines #
#   J1, J2, J3, and L1, should be viewed as an intertwined group. They should #
#   not be viewed in a strictly sequential manner. Throughout most of SWMM #
#   the user can count on one line leading logically to the next or perhaps #
#   negating some subsequent lines or portions thereof. Here the user will encounter #
#   cases in which he or she will find that something on an earlier line is negated #
#   or rendered irrelevant by a later line! Beware!             #
##########################################################################
```

*J1* : *GENERAL QUALITY CONTROL DATA*

NQS : Number of quality constituents (maximum of 10).

****************************************************
* It is assumed here that erosion will not be simulated, but if it is*
* the last constituent is reserved for the eroded solids. *
****************************************************

JLAND : Number of land uses (maximum of 5).

****************************************************
* JLAND land use types are developed in the J2 lines and *
* used to characterize the land use for each subcatchment *
* described in the L1 lines. *
****************************************************

IROS : Erosion simulation parameter.
= 0, erosion not simulated.
= 1, erosion of suspended solids simulated using the Universal Soil Loss Equation.

****************************************************
* As mentioned above the simulation of erosion is ignored in *
* this guide, i.e., set IROS = 0. *
****************************************************

IROSAD : Option to add erosion constituent to constituent number IROSAD. No addition if IROSAD = 0. Not required if IROS = 0.

****************************************************
* As mentioned above the simulation of erosion is ignored *
* in this guide, i.e., set IROSAD = 0. *
****************************************************

DRYDAY : Number of dry days prior to start of simulation.

****************************************************
* This is used in the pollutant buildup algorithm to establish *
* a surface load prior to the simulation. It has little or no effect *
* on long-term, continuous simulations, but can have an *
* important effect on single-event runs. *
****************************************************

CBVOL : Average individual catchbasin storage volume, cubic feet [cubic meters].

****************************************************
* Catchbasins usually do not have a major impact on runoff *
* quality (with the possible exception of floatables). Little *
* effort should be spent to characterize them. If catchbasins *
* are not simulated, set CBVOL = 0.0. *
****************************************************

DRYBSN : Dry days required to recharge catchbasin concentrations to initial values (CBFACT on line J3). DRYBSN must be > 0.

```
**************************************************
* See comments for CBVOL. In any case, set DRYBSN > 0. *
* If CBVOL = 0, the value of DRYBSN is irrelevant.     *
**************************************************
```

RAINIT : Highest average 30-minute rainfall intensity during the year (continuous SWMM) or during the storm (single event), inches/hour [millimeters/hour]. Not required if IROS = 0.

```
**************************************************
* As mentioned above, the simulation of erosion is ignored, *
* i.e., set RAINIT = 0.0.                                  *
**************************************************
```

```
*****************************************************************
*   The next three parameters are used to model street sweeping. KLNBGN and  *
*   KLNEND are only used if the simulation is greater than 1 month.          *
*                                                                            *
*   Generally, only the most vigorous street sweeping programs have          *
*   significant impact on runoff quality.                                    *
*****************************************************************
```

REFFDD : Street sweeping efficiency (removal fraction) for "dust and dirt."

```
**************************************************
* This may or may not be relevant depending on whether any *
* constituents are related to "dust and dirt". This is determined*
* in the J3 line(s).                                         *
**************************************************
```

KLNBGN : Day of year on which street sweeping begins (e.g., March 1 = 60).

KLNEND : Day of year on which street sweeping stops (e.g., November 30 = 334).

---

*J2 LINES : LAND USE DATA*

---

```
*****************************************************************
*   Each of the JLAND land uses (JLAND was entered in line J1) is described   *
*   here with a different J2 line. The order of entry determines the number assigned *
*   to each land use, i.e., land use 1 will be represented by the first J2 line, land *
*   use 2 will be represented by the second J2 line, etc.                      *
*****************************************************************
```

LNAME(J) : Name of land use J.

**************************************************
* Give the land use a name in single quotes, *
* e.g., 'RESIDENTIAL'. *
**************************************************

METHOD (J) : Buildup equation type for "dust and dirt" build up as a function of time on land use J.
= 0, power-linear equation,
= 1, exponential equation,
= 2, Michaelis-Menton equation.

**************************************************
* These equations are discussed in Chapter 2 and in great *
* detail in the SWMM User's Manual. There is no clear choice,*
* but it may be fair to say that the differences they produce *
* in terms of simulated runoff quality are small compared *
* to the errors and uncertainty inherent in the model *
* and water quality measurements. Pick one and use it *
* consistently. If time and resources permit, some experi- *
* mentation may be marginally useful. *
* *
* One of the truly confusing things here is that constituent *
* buildup need not be related to "dust and dirt". As seen in line *
* J3 (and Table 4-18 in the SWMM User's Manual), constituent *
* buildup can be simulated by avoiding the "middleman" of *
* dust and dirt and relating it directly to curb length or area. *
* If no constituents are related to "dust and dirt", METHOD is *
* irrelevant; but if one or more are related to "dust and dirt", *
* METHOD is relevant. *
* *
* <u>This is a "default/ratio" indicator</u>. The parameters below with *
* asterisks are those to which this option may be applied. See the*
* discussion under "A.1 General Comments" at the beginning *
* of this Appendix. *
**************************************************

JACGUT (J) : Functional dependence of build up parameters for use in the "dust and dirt" build up equation for land use J.
= 0, function of subcatchment gutter (curb) length,
= 1, function of subcatchment area.
= 2, constant.

**************************************************
* Dust and dirt buildup can be related to gutter length, i.e., *
* pounds/100 feet of curb [kilograms/kilometer of curb], *
* subcatchment area, i.e., pounds/acre [kilograms/hectare], *
* or neither (what's called "constant" above), i.e., pounds or *
* kilograms. In all cases, the buildup is a function of time. *
**************************************************

```
************************************************************************
*    The following three entries are the parameters for the "dust and dirt" buildup  *
*    equation used for land use J. There may be as many as three parameters depending *
*    on the equation used. (See Table 4-16 in the SWMM User's Manual).                *
*                                                                                      *
*    Be careful using Tables 4-16, 4-17, and 4-18. They are very difficult to         *
*    understand but with some effort they can be mastered. Spend some time linking    *
*    them together.                                                                    *
************************************************************************
```

DDLIM(J)* : Limiting buildup quantity.

DDPOW(J)* : Power or exponent.

DDFACT(J)* : Coefficient.

```
************************************************************************
*    The following three entries are the street sweeping parameters for land          *
*     use J.                                                                           *
*                                                                                      *
*    Again, only the most vigorous street sweeping programs have any real impact on   *
*    runoff quality. The amount of input data is minimal, though, and the user        *
*    may wish to include it. However, if street sweeping is not simulated, simply set *
*    all of these parameters to zero.                                                  *
*                                                                                      *
*    It is not necessary for the "gutter length" option to be used, i.e., JACGUT = 0, *
*    for street sweeping to be simulated. Street sweeping is not related to "dust     *
*    and dirt". Street sweeping is simulated by reducing the available constituent    *
*    surface load regardless of how the load is calculated. The term "available"      *
*    refers to how much of the load is actually accessible by the sweeper.            *
************************************************************************
```

CLFREQ(J)* : Cleaning interval, days.

```
*******************************************************
* Sweeping will not have a significant impact if the cleaning *
* interval is much longer than the mean interevent time. Intervals *
* of a month or longer will probably have no significant effect. *
*******************************************************
```

AVSWP(J)* : Availability factor, fraction.

```
*******************************************************
* The parameter AVSWP represents the surface load available*
* for sweeping. It is applied to all quality constituents.    *
*******************************************************
```

DSLCL(J)* : Days since the last cleaning before the start of the simulation. DSLCL should be ≤ CLFREQ.

**************************************************
* This is used to determine how much of the surface load was*
* removed by sweeping before the simulation begins. *
**************************************************

*J3 LINES : QUALITY CONSTITUENT(S) DATA*

****************************************************************
* Each of the NQS quality constituents (NQS was entered in line J1) is described *
* here with a different J3 line. The order of entry determines the identification *
* number assigned to each constituent, i.e., constituent 1 will be represented *
* by the first J3 line, constituent 2 will be represented by the second J3 line, etc. *
****************************************************************

PNAME(K) : Constituent name.

**************************************************
* Give constituent K a name in single quotes, e.g., 'SUSPENDED *
* SOLIDS'. *
**************************************************

PUNIT(K) : Name of concentration units for constituent K.

**************************************************
* Give constituent K units a name in single quotes, e.g., *
* 'MG/L'. The units are actually specified immediately below.*
**************************************************

NDIM(K) : Type of concentration units for constituent K.
= 0, mass per liter, mg/L.
= 1, "other" quantity per liter, e.g., MPN/L.
= 2, other concentration units, e.g., pH, JTU.

**************************************************
* This essentially makes sure that the proper conversion factors *
* are used in the program. Of course, the units associated with*
* NDIM = 0 are the most common. *
**************************************************

KALC(K) : Type of buildup calculation for constituent K.
= 0, buildup is fraction of "dust and dirt" for each land use.
= 1, power-linear equation.
= 2, exponential equation.
= 3, Michaelis-Menton equation.
= 4, no buildup required (KWASH = 1 below).

**************************************************
* Again, note that the "dust and dirt" parameters in lines J1 *
* and J2 are irrelevant if KALC = 1, 2, 3, or 4. As with the *

```
* "dust and dirt" buildup equations, there is no clear choice.  *
* Similarly, it is likely that the differences they produce in  *
* terms of simulated water quality are small compared to the    *
* errors and uncertainty inherent in the model and water quality *
* measurements. Pick one and use it consistently. If time and   *
* resources permit some experimentation may be marginally       *
* useful.                                                       *
*                                                               *
* The user should note, however, the difference between the     *
* "dust and dirt"/constituent buildup method and the constituent *
* buildup method. The former allows constituent buildup to be a *
* function of constituent behavior and land use (since "dust and*
* dirt" buildup is characterized for each land use in line J2). *
* For the latter, constituent buildup is only a function of     *
* constituent behavior.                                         *
*                                                               *
* This is a "default/ratio" indicator. The parameters below with *
* asterisks are those to which this option may be applied. See the*
* discussion under "A.1 General Comments" at the beginning      *
* of this Appendix.                                             *
*****************************************************************
```

KWASH(K) : Type of washoff calculation for constituent K.
= 0, power-exponential washoff equation.
= 1, rating curve (loading rate is a function of runoff only), no upper limit.
= 2, rating curve, upper limit is calculated by the buildup equation (KALC = 0, 1, 2, or 3).

```
*****************************************************************
* If one has gone through the trouble of estimating the         *
* buildup  of surface loads KWASH should probably equal         *
* 0. In other words, the accumulated load will be washed off    *
* the surface with a  fairly well known relationship. If        *
* KWASH  = 1, all the buildup calculations are irrelevant       *
* since the delivered loading rate is simply a function of the  *
* runoff flow rate. If KWASH = 2, the delivered loading rate    *
* is calculated as with KWASH = 1. However, the total delivered*
* load will not be allowed to exceed the accumulated constituent *
* surface load (so the buildup calculations become somewhat     *
* relevant).                                                    *
*****************************************************************
```

KACGUT(K) : Functional dependence of buildup parameters in the buildup equation for constituent K. Not required for KALC = 0 or 4.
= 0, function of subcatchment gutter (curb) length.
= 1, function of subcatchment area.
= 2, constant.

```
************************************************************
* The parameters in the build equations are related to curb   *
* length or area or neither (i.e., "constant"). Refer carefully *
* to Tables 4-17 and 4-18 in the SWMM User's Manual.          *
************************************************************
```

LINKUP(K) : Linkage to snowmelt. Not required if ISNOW = 0 (line B1) or KALC = 4.

- = 0, no linkage to snow parameters.
- = 1, constituent buildup during dry weather only when snow is present on impervious surface of subcatchment.

```
************************************************************
* Snowmelt is ignored in this guide. Set LINKUP = 0.          *
************************************************************
```

```
****************************************************************************
* The following items are the parameters for the buildup equation used for *
* constituent K. There may be as many as five parameters depending on the  *
* equation used. See Tables 4-17 and 4-18 in the SWMM User's Manual.       *
*                                                                          *
* The user is again reminded to be very careful using Tables 4-16, 4-17, and *
* 4-18. They are very difficult to understand but with some effort they can be *
* mastered. Spend some time linking them together.                         *
****************************************************************************
```

QFACT(1,K)* : First buildup parameter, e.g., limit.

QFACT(2,K)* : Second buildup parameter, e.g., power or exponent.

QFACT(3,K)* : Third buildup parameter, e.g., coefficient.

QFACT(4,K)* : Fourth buildup parameter. Not required if KALC > 0.

QFACT(5,K)* : Fifth buildup parameter. Not required if KALC > 0.

```
****************************************************************************
* The following two items are the washoff (for KWASH = 0) or rating curve  *
* (for KWASH = 1 or 2) parameters for constituent K.                       *
****************************************************************************
```

WASHPO(K)* : Power (exponent) for the runoff rate.

RCOEF(K)* : Coefficient.

CBFACT(K)* : Initial catchbasin concentration (units according to NDIM).

```
************************************************************
* See comments for CBVOL, line J1.                            *
************************************************************
```

CONCRN(K)* : Concentration in precipitation (units according to NDIM).

```
*******************************************************
* Precipitation loads can have some impact, but few users     *
* will have the data to substantiate a number here. Since     *
* measured runoff quality will include the precipitation      *
* contribution it is probably better, in the absence of rainfall *
* quality data, to set CONCRN = 0 and assume that the         *
* contribution is indirectly included when calibrating the buildup *
* and washoff parameters.                                     *
*******************************************************
```

REFF(K)* : Street sweeping efficiency (removal fraction) for this constituent.

```
*******************************************************
* The availability of all constituents (i.e., the fraction of all *
* loads that can be reached by the sweepers) for a given land use*
* was specified in line J2. This parameter sets the efficiency *
* of the sweeper at picking up constituent K.                 *
*******************************************************
```

*J4 LINE :CONTRIBUTIONS FROM OTHER CONSTITUENTS*

```
*******************************************************************
* Line J4 allows a fraction of one constituent to be added to another. This is rarely*
* used and it is omitted here.                                    *
************* *****************************************************
```

*K1 LINE : EROSION DATA*

```
*******************************************************************
* Line K1 is omitted because the erosion option was not selected in line J1, *
* i.e., IROS = 0.                                                 *
*******************************************************************
```

*L1 LINES : SUBCATCHMENT SURFACE QUALITY DATA*

```
*******************************************************************
* The L1 lines are used to assign land use type and other characteristics to *
* each subcatchment. One line is required for each subcatchment. The order of *
* entry is arbitrary, but a match, via the subcatchment or name (NAMEW), must *
* be found in the H1 lines.                                       *
*******************************************************************
```

K=NAMEW : Subcatchment number or name. There must be an exact match in the H1 lines.

```
****************************************************
* This is a "default/ratio" indicator. The parameters below    *
* with asterisks are those to which this option may be applied. See *
* the discussion under "A.1 General Comments" at the           *
* beginning of this Appendix.                                    *
****************************************************
```

KL : Land use classification for subcatchment N, numbers 1 through 5 corresponding to the input sequence in the J2 lines.

BASINS(N)* : Number of catchbasins in subcatchment N.

```
****************************************************
* See comments for CBVOL, line J1.                               *
****************************************************
```

GQLEN(N)* : Total curb length in subcatchment N, hundreds of feet [kilometers]. May not be required depending on the method used to calculate constituent loadings (see lines J2 and J3.

```
****************************************************
* This is necessary only if the "dust and dirt" or constituent   *
* buildup is related to the curb or gutter length in the subcatchment,*
* i.e., JACGUT = 0 (line J2) or KACGUT = 0 (line J3).             *
****************************************************
```

```
*****************************************************************
*   The following initial surface loading values may be input as an alternative to the  *
*   computation of loads via the methods specified in lines J2 and J3. Each              *
*   non-zero value is multiplied by the subcatchment area to produce the total           *
*   initial load for the subcatchment. The units for that load depend on the value of   *
*   NDIM (line J3), according to the following table:                                    *
*                                                                                        *
*            NDIM        LOAD                                                            *
*                                                                                        *
*              0       pounds [kilograms]                                                *
*              1       10^6 · quantity, e.g., 10^6 MPN                                   *
*              2       10^6 · quantity · cubic feet, e.g., 10^6 pH-cubic feet.           *
*                                                                                        *
*   The order of entry is the same as that found in the J3 lines. Entries must be        *
*   made for 10 constituents (even if NQS < 10). These can, of course, be zeroes.        *
*                                                                                        *
*   If more than one event is simulated the buildup data on lines J2 and J3 will         *
*   still be needed to recharge the surface loads after the first event. Therefore,      *
*   it makes little sense to use this option for anything but single event               *
*   simulation.                                                                          *
*****************************************************************
```

PSHED(1,N) : Initial surface loading, constituent 1, load/acre (load/hectare).

. .

. .

. .

PSHED(10,N) : Initial surface loading, constituent 10, load/acre (load/hectare).

---

*M1 LINE : CHANNEL/INLET PRINT CONTROL DATA*

****************************************************************
* Since it was assumed that all drainage system routing will be done with the *
* TRANSPORT or EXTRAN Blocks, only flows and concentrations at inlets will be *
* printed. Each inlet receives the discharge from one or more subcatchments. *
****************************************************************

NPRNT : Total number of channels/pipes/inlets for which non-zero flows (and concentrations, if simulated) are to be printed.

************************************************
* The maximum number varies from one SWMM release *
* to another. *
************************************************

INTERV : Print control.
= 0, print statistical summary only.
= 1, print every time step.
= K, print every K time steps.

************************************************
* If INTERV = 0, line M2 becomes irrelevant and no detailed *
* printout will be created for any period. *
************************************************

---

*M2 LINE : DETAILED PRINT PERIOD INFORMATION*

****************************************************************
* IF NPRNT = 0 ON LINE M1, OMIT THIS LINE. THIS LINE BECOMES *
* IRRELEVANT IF INTERV = 0 IN LINE M1, BUT IT MUST BE *
* INCLUDED (SIMPLY ENTER ZERO FOR EACH OF THE ITEMS). *
* *
* This line allows the user to specify the time periods over which detailed *
* output is to be printed for various inlets (the level of detail is indicated by *
* NPRNT = 1 or K in line M1; inlets for which this is done are indicated in *
* line M3). If the simulation time is long, print one or two short periods covering *
* a few storm events to check the output. Each printed interval at a given location *
* will produce at least one line of output, so beware. *
* *

```
*   If NDET = 1 and STARTP(1) = 0 and STOPPR(1) = 0, the total simulation   *
*   period will be printed.                                                 *
*                                                                           *
*   Table 4-29 in the SWMM User's Manual presents a nice summary of the     *
*   output options.                                                         *
*****************************************************************************
```

NDET : Number of detailed printout periods (maximum of 10 periods.)

STARTP(1) : First starting printout date, year/month/day (e.g., October 2, 1949 = 491002).

STOPPR(1) : First stopping printout date.

. .

. .

. .

STARTP(NDET) : Last starting printout date.

STOPPR(NDET) : Last stopping printout date.

---

*M3 LINE : CHANNEL/INLET PRINTOUT LOCATIONS*

```
*****************************************************************************
*   OMIT THIS LINE IF NPRNT = 0 ON LINE M1.                                 *
*                                                                           *
*   Since it was assumed that all drainage system routing will be done with the *
*   TRANSPORT or EXTRAN Blocks, only flows and concentrations at inlets     *
*   will be printed. Each inlet receives the discharge from one or more     *
*   subcatchments.                                                          *
*****************************************************************************
```

IPRNT(1) : First channel/inlet number or name for which flows and concentrations are to be printed.

. .

. .

. .

IPRNT(NPRNT): Last channel/inlet number or name for which flows and concentrations are to be printed.

## A.3 TRANSPORT BLOCK INPUT DATA

See NOTE at beginning of Section A.2.

---

*A1 LINE : TITLE LINES*

TITLE : Two lines with heading to be printed on output.

```
*****************************************************
* There must be two lines, each in single quotes ('Run No. 1'). *
*****************************************************
```

*B0 LINE : MISCELLANEOUS CONTROL PARAMETERS*

*****************************************************************
* The latest versions of TRANSPORT contain a B0 line with some options that do *
* little to improve model performance. The line may be omitted and so it is *
* here. *
*****************************************************************

*B1 LINE : CONTROL DATA*

*****************************************************************
* This line sets some of the run control parameters and TRANSPORT Block *
* options, as do lines B2 and B3. *
*****************************************************************

NDT : Number of time steps.

*****************************************************
* There is no limit on the number of time steps. However, *
* the TRANSPORT Block is definitely not designed for long- *
* term simulation. Daily and monthly summaries are not *
* printed and zero values are not suppressed. *
* *
* The total simulation time will be NDT · DT. DT is entered in *
* line B2. The simulation time will be compared to the time from the *
* preceding block (usaully RUNOFF). The lesser of the two will *
* be used to determine the length of the TRANSPORT *
* Block simulation. *
*****************************************************

NINPUT : Number of non-conduit elements with input of hydrographs and pollutographs in the R1 lines. The maximum number of locations is 80.

*****************************************************
* Hydrographs and pollutographs are seldom entered with the *
* R1 lines. This option will not be used here. Set NINPUT = 0. *
*****************************************************

NNYN : Number of non-conduit elements for which input hydrographs and pollutographs are printed. The locations are entered in the J1 line. The maximum number of locations is 80.

*****************************************************
* "Input" usually refers to data transferred from an earlier *
* block, e.g., RUNOFF Block. This can be useful in checking *
* for the correct transfer of data. *
*****************************************************

NNPE : Number of non-conduit elements at which routed hydrographs and pollutographs are printed. The locations are entered in the J2 line. The maximum number of locations is 80.

```
*****************************************************
* The routed hydrographs and pollutographs include flows   *
* and pollutants from upstream elements and those entered  *
* at that element.                                          *
*****************************************************
```

NOUTS : Number of non-conduit elements from which routed hydrographs and pollutographs are placed on an interface file for transfer to subsequent block. The locations are entered in the I1 line. The maximum number of locations is 100.

```
*****************************************************
* The "subsequent" block is most likely the STORAGE/        *
* TREATMENT Block (which is not covered in this guide) or*
* a receiving water model.                                  *
*****************************************************
```

NPRINT : Status of error messages.
= 0, suppress error messages.
= 1, print up to 100 error messages.

```
*****************************************************
* These error messages indicate non-convergence problems in *
* the flow routing algorithm. The numerical algorithms in   *
* TRANSPORT are usually very well behaved and there will *
* few, if any, of these messages in most cases. However, the *
* user should be aware of any convergence problems. Set     *
* NPRINT = 1.                                               *
*****************************************************
```

NPOLL : Number of water quality constituents (maximum of four). When NPOLL = 0 all water quality operations will be ignored. The constituents are specified in the F1 lines.

```
*****************************************************
* Up to four water quality constituents may be transferred *
* from an earlier block, e.g., RUNOFF Block.                *
*****************************************************
```

NITER : Number of iterations used in routing subroutine.

```
*****************************************************
* There is an algorithm in the numerical flow routing routine *
* that controls unwanted flow oscillations. The number of    *
* iterations used by this algorithm is entered here. Four    *
* iterations will nearly always be adequate, i.e., set NITER = 4.*
*****************************************************
```

IDATEZ : Starting date of storm, year/month/day, e.g., July 20, 1979 = 790720. This value is superseded by value transferred from preceding block.

METRIC : Metric input-output.
= 0, use U.S. customary units.
= 1, use metric units. Metric input is indicated in brackets [ ] through the remainder of this guide.

INTPRT : Interval for printing of input and output data at the selected elements, i.e., data will be printed for each selected element every INTPRT time steps. If INTPRT = 0, only total flows and loads and some basic statistics will be printed.

---

*B2 LINE : MORE CONTROL DATA*

DT : Time step size, seconds.

```
****************************************************
* The TRANSPORT Block is not particularly sensitive to      *
* the size of the time step. Times of 5, 10, and 15 minutes  *
* or so will usually work well. Very short times of 60 seconds *
* or less are usually unnecessary. However, the time step    *
* should reflect flow conditions. If rapid changes in flow are *
* expected, the time step should be set accordingly.         *
*                                                            *
* The value of DT does not have to agree with the time step  *
* transferred from the RUNOFF Block (in fact this would      *
* be difficult since RUNOFF can have variable time steps).   *
* TRANSPORT will estimate flows and concentrations from      *
* the interface file by interpolation. However, the interpolation *
* algorithm works best when the time specified here is an    *
* integer multiple (e.g., 2 or 3 times) or integer fraction (e.g., *
* 1/2 or 1/3) of the time step from the preceding block.     *
* If variable time steps are encountered use the shortest to *
* calculate the integer multiple or fraction.                *
****************************************************
```

EPSIL : Allowable error for convergence of the iterative methods in the flow routing routine. A value of 0.0001 is recommended.

```
****************************************************
* Use the recommended value.                                 *
****************************************************
```

DWDAYS : Total number of dry weather days prior to the start of the simulation during which solids were not flushed from the sewers.

**************************************************

* This parameter is used to compute the amount of solids settled *
* in the sewers prior to the start of the simulation. In this *
* guide the settling and scour of solids from the sewers is *
* ignored (see line F1). <u>Set DWDAYS = 0</u>. *

**************************************************

TZERO : Starting time of the simulation run expressed as hours from midnight, e.g., 5:30 p.m. is entered as 17.5. This value will be superseded by the starting time transferred from a preceding block.

GNU : Kinematic viscosity of water, square feet/second [square meters/second]. Used if SPG > 1.0 for constituents on line F1.

**************************************************

* This parameter is used in the algorithm to simulate the *
* settling and scour of solids in the sewers. In this guide this *
* option is not considered (see line F1). <u>Set GNU = 0.0</u>. *

**************************************************

TRIBA : Total catchment area, acres [hectares]. This value is superseded by the value transferred from the preceding block.

---

*B3 LINE : MORE CONTROL DATA*

---

NCNTRL : Control parameter specifying the means in obtaining the input hydrographs and pollutographs.
= <u>0</u>, input hydrographs and pollutographs obtained from a preceding block via an interface file.
= 1, input obtained from R1 lines.

**************************************************

* It will be assumed here that the input will come from a *
* preceding block, i.e., the RUNOFF Block. Entering flow and *
* quality constituent data with the R1 lines is very awkward. *
* There are potential uses for the R1 lines, but this option is *
* seldom used. <u>Set NCNTRL = 0</u>. *

**************************************************

NINFIL : Control parameter to determine if sewer infiltration will be simulated.
= 0, sewer infiltration is not estimated.
= 1, sewer infiltration is estimated.

**************************************************

* The information required for the infiltration routine is entered *
* with lines K1 and K2. This option is not particularly useful. *
* Estimates for infiltration can be entered directly at the non- *
* conduits or nodes (see E1 line). *

**************************************************

NFILTH : Control parameter to determine if sanitary sewage flows and quality constituent loads will be estimated.
= 0, sewage flows not estimated.
= 1, sewage flows are estimated.

```
****************************************************
* The information required to estimate sanitary sewage is    *
* entered in lines L1, L2, L3, M1, M2, M3, M4, N1, O1, O2, *
* P1, and P2. The amount of information needed is large and the *
* usefulness marginal. This option is not considered here (see *
* the discussion preceding Line L1). Set NFILTH = 0. Estimates*
* for dry weather flow can be entered directly at the non-    *
* conduits or nodes (see E1 line).                            *
****************************************************
```

NDESN : Control parameter to determine if the hydraulic design routine will be used.
= 0, hydraulic design routine is not used.
= 1, hydraulic design routine is used.

```
****************************************************
* This routine adjusts conduits that are not capable of passing *
* the flow encountered during the simulation without          *
* surcharging. While the routine functions adequately it is   *
* probably better for the user to handle this directly. NDESN *
* = 0 is suggested.                                           *
****************************************************
```

---

```
#####################################################################
#  Lines C1, D1, and D2 are used to develop the hydraulic characteristics for  #
#  user-supplied conduit shapes (i.e., shapes not included in the TRANSPORT  #
#  Block menu — see the E1 line below). User-supplied conduit shapes are      #
#  seldom needed. Thus, this capability is ignored here. However, line C1 must #
#  be included.                                                               #
#####################################################################
```

---

*C1 LINE : NUMBER OF USER-SUPPLIED CONDUIT SHAPES*

---

NKLASS : Number of user-supplied conduit shapes. Set NKLASS = 0.

KPRINT : Control parameter for printing flow routing parameters for all shapes (program and user supplied).
= 0, suppress printing.
= 1, allow printing (about 500 lines).

---

*D1 LINE : NAMES OF USER-SUPPLIED CONDUIT SHAPES*

---

*D2 LINE : NUMBER OF VALUES FOR DNORM*

---

*E1 LINES : SEWER ELEMENT DATA*

---

********************************************************************
* This line is repeated for each element (non-conduit or conduit) in the modeled*
* sewer system. There is a limit on the number of elements (which varies from one *
* SWMM release to another), but it may be increased (see the SWMM User's *
* Manual). *
********************************************************************

NOE : Element number. This may be a positive integer as long as 9 digits.

**************************************************
* These numbers are essentially arbitrary. The numbers *
* themselves do no indicate any ordering or networking (e.g., *
* element 5 is not necessarily connected to element 6). The *
* sewer network is developed as a cascade of elements, with *
* the elements upstream of element NOE specified below. An *
* internal numbering scheme uses this information to develop *
* the proper network. *
* *
* <u>This is a "default/ratio" indicator</u>. The parameters below *
* with asterisks are those to which this option may be applied.*
* See the discussion under "A.1 General Comments" at the *
* beginning of this Appendix. *
**************************************************

********************************************************************
* Up to three elements may be placed upstream of element NOE. They are *
* entered by placing upstream element numbers (NOE in the particular E1 line *
* describing that element) in one or more of the slots below. Enter zeros *
* for unused upstream element slots. *
********************************************************************

NUE(1) : First of three possible upstream elements.

NUE(2) : Second of three possible upstream elements.

NUE(3) : Third of three possible upstream elements.

********************************************************************
* The remaining parameters on this line describe the element. *
* The element type is identified with NTYPE. *
* *
* More specific characteristics are entered with the remaining parameters. Not *
* all of these parameters are needed for each element type and the same *
* parameter has different meanings for different types (e.g., GEOM1 = *
* diameter when NTYPE = 1, GEOM1 = weir height when NTYPE = 23). *
* The user MUST consult Tables 6-3 and 6-4 and Figures 6-3 and 6-4 in the *
* SWMM User's Manual for the meaning of each parameter. However, the *
* meaning of each parameter for a conduit or a manhole (the most common *
* non-conduit) is indicated. *
********************************************************************

NTYPE : Element type.

Conduits

= 1, circular.
= 2, rectangular.
= 3, egg-shaped.
= 4, Boston horseshoe.
= 5, Gothic.
= 6, caternary.
= 7, semi-elliptic.
= 8, basket-handle.
= 9, semicircular.
= 10, modified basket-handle.
= 11, rectangular, triangular bottom.
= 12, rectangular, round bottom.
= 13, trapezoidal channel.
= 14, parabolic channel.
= 15, power function channel.
= 16, natural (irregular) channel.
= 17,18, user supplied.

```
*****************************************************
* Conduit options 16, 17, and 18 are not addressed here.      *
* They are rarely needed and involve extensive data entry.    *
* Most conduits can be modeled with options 1 through 15.     *
*****************************************************
```

Non-conduits

= 19, manhole.
= 20, lift station.
= 21, flow divider, simple diversion structure.
= 22, internal storage unit. (This option allows the placement of storage units throughout the sewer system. It requires additional input in the G1 through G5 lines.)
= 23, flow divider, weir diversion structure.
= 24, flow divider, cunnette section.
= 25, backwater element.

```
*****************************************************
* The TRANSPORT Block has a clumsy method of modeling *
* backwater effects (using option 25, backwater element). If *
* backwater is a serious consideration, use the EXTRAN      *
* Block!                                                    *
*****************************************************
```

DIST* : Element length for conduit, feet [meters].

OR

Constant flow to a manhole, cubic feet/second [cubic meters/second].

```
*****************************************************
* Constant flows and concentrations may only be entered at  *
* manholes (NTYPE = 19).                                    *
*****************************************************
```

GEOM1* : First characteristic dimension of the conduit, feet [meters].

OR

Constant concentration of quality constituent 1 in the inflow entered with DIST, if water quality routing is simulated (NPOLL ≥ 1 in line B1). The units should agree with those specified with NDIM in line F1.

SLOPE* : Invert slope of the conduit, percent or feet/100 feet [meters/100 meters].

OR

Constant concentration of quality constituent 2 in the inflow entered with DIST, if water quality routing is simulated and more than one constituent is specified (NPOLL ≥ 2 in line B1). The units should agree with those specified with NDIM in line F1.

ROUGH* : Manning's roughness of the conduit.

OR

Constant concentration of quality constituent 3 in the inflow entered with DIST, if water quality routing is simulated and more than one constituent is specified (NPOLL ≥ 3 in line B1). The units should agree with those specified with NDIM in line F1.

GEOM2* : Second characteristic dimension of the conduit.

OR

Constant concentration of quality constituent 4 in the inflow entered with DIST, if water quality routing is simulated and more than one constituent is specified (NPOLL ≥ 4 in line B1). The units should agree with those specified with NDIM in line F1.

BARREL : Number of barrels for the conduit. The barrels are assumed to be identical in shape and flow characteristics. (Non-conduits do not use this parameter.)

GEOM3* : Third characteristic dimension of the conduit, feet [meters]. (Manholes do not use this parameter, but some other non-conduits do.)

KGEOM : Third characteristic dimension of the conduit if alphanumeric elements are used.

*******************************************************
* This parameter is ignored here. Set KGEOM = ‘ ’. *
*******************************************************

######################################################################
# The E2, E3, and E4 lines are used to enter hydraulic data for natural channels #
# (NTYPE = 16 in line E1). This option is rarely used and, thus, these lines #
# are omitted here. #
######################################################################

*E2 LINE : NATURAL CHANNEL ROUGHNESS DATA*

*E3 LINE : NATURAL CHANNEL CROSS SECTION DATA*

*E4 LINE : NATURAL CHANNEL PROFILE DATA*

*F1 LINES : WATER QUALITY DATA*

*******************************************************************
* OMIT THIS LINE IF NPOLL = 0 IN LINE B1. *
* *
* This line is used to enter water quality data and is repeated for each of NPOLL *
* constituents (maximum of four, line B1). *
*******************************************************************

KPOL : Water quality constituent selected from interface file. If KPOL = 0, this constituent is not read from the interface file, but is defined below.

*******************************************************
* The "interface" file will, of course, almost always come from *
* the RUNOFF Block. The number entered here should *
* correspond to the order in which the constituent of interest *
* appears on the interface file. For the RUNOFF Block, the *
* order on the interface file is identical to that of J3 lines in the *
* RUNOFF data. For example, if BOD was the fifth constituent *
* in the RUNOFF Block, set KPOL = 5. *
*******************************************************

*******************************************************************
* The next three parameters are not required if the constituent was taken from *
* the interface file. When KPOL > 0 the information requested with these *
* parameters is transferred from the interface file.
*******************************************************************

PNAME : Constituent label, if KPOL = 0.

PUNIT : Constituent units, if KPOL = 0.

NDIM : Type of concentration units for constituent K, if KPOL = 0.
= 0, mass per liter, mg/L.
= 1, "other" quantity per liter, e.g., MPN/L.
= 2, other concentration units, e.g., pH, JTU.

```
****************************************************
*  This essentially ensures that the proper conversion factors  *
*  are used in the program. Of course, NDIM = 0 is the          *
*  most common.                                                 *
****************************************************
```

DECAY : First-order decay coefficient, 1/day.

```
****************************************************
*  A completely mixed flow regime is assumed to occur in each  *
*  conduit and in internal storage units (NTYPE = 22) and      *
*  each constituent is assumed to behave as a first-order      *
*  reactant. If the constituent is conservative, set DECAY= 0.0.*
*                                                              *
*  Travel times are relatively short in sewer systems and therefore*
*  decay can probably be ignored. However, decay will          *
*  probably be significant in storage units.                   *
****************************************************
```

```
*********************************************************************
*   The remaining parameters on this line allow the user to assign a particle size  *
*   distribution to each constituent. This information is used in the simulation of *
*   scour and deposition in the sewer system. These phenomena certainly occur       *
*   in sewer systems, but their magnitudes are usually unknown. It is assumed       *
*   through the rest of this guide that deposition/scour will not be modeled, i.e., *
*   set SPG = 0.0. In addition set the remaining parameters to 0.0.                 *
*********************************************************************
```

SPG : Specific gravity. If SPG > 1.0 scour/deposition is modeled.

PSIZE(2) : Second particle size, millimeters. PSIZE(1) is automatically set to 0.0 millimeters.

PGR(2) : Percent greater than PSIZE(2). PGR(1) is automatically set to 100%.

PSIZE(3) : Third particle size, millimeters.

PGR(3) : Percent greater than PSIZE(3).

PSIZE(4) : Fourth particle size, millimeters.

PGR(4) : Percent greater than PSIZE(4).

PSIZE(5) : Fifth particle size, millimeters.

PGR(5) : Percent greater than PSIZE(5).

PSDWF : Maximum particle size contained in dry-weather flow input. PSDWF must be ≤ PSIZE(5).

```
##########################################################################
#   OMIT LINES G1, G2, G3, G4, AND G5 IF NO INTERNAL STORAGE             #
#   ELEMENTS ARE SPECIFIED IN THE E1 LINES.                              #
#                                                                        #
#   This group of lines is repeated for each element identified in the E1 lines as  #
#   an internal storage element (although not every line may be necessary for a     #
#   given element).                                                      #
#                                                                        #
#   The internal storage algorithm gives storage attributes to non-conduits (nodes)#
#   rather than modeling them as simple "flow in" = "flow out" devices. This is    #
#   a very useful tool for investigating the effects of in-line storage without    #
#   having to resort to the rather cumbersome task of using the STORAGE/           #
#   TREATMENT Block.                                                     #
#                                                                        #
#   The order of entries here should correspond to the order in which internal    #
#   storage elements appear in the E1 lines, i.e., the first storage element in    #
#   the E1 lines (regardless of how many other elements were listed before it)     #
#   should be the focus of the first group of G1-G5 lines.               #
##########################################################################
```

---

*G1 LINES : INTERNAL STORAGE ELEMENTS*

---

LOUT(IS) : Outflow routing parameter for internal storage element IS.

= 0, depth-outflow relationship is described by as many as 16 pairs in line G2.

= 1, depth-outflow relationship is described by a single power equation in line G3.

= 2, depth-outflow relationship is described by two power equations in line G3.

= 3, depth-outflow relationship is controlled by pumps described in line G4.

```
*****************************************************
* This parameter selects the method by which the depth–     *
* outflow relationship for the storage element will be specified. *
* LOUT = 0 provides the most versatility.                    *
*****************************************************
```

---

*G2 LINES : DEPTH-SURFACE AREA-VOLUME-OUTFLOW DATA*

---

```
**************************************************************************
*   The G2 line is used to provide the surface area-volume-outflow information for  *
*   a particular depth. Repeating this line for several depths allows the hydraulic *
*   behavior of the unit to be estimated.                                 *
*                                                                         *
```

* The lines are ordered from the bottom of the unit (i.e., where depth or TSDEP = 0.0). As many as 16 lines (i.e., depths) may be used to characterize the unit. The volumes (i.e., TSTORE) may be replaced with zeroes, in which case volume at each depth is estimated by the program with the depth-surface area data. The outflow may be omitted if the depth-outflow relationship is specified by a power equation (an option indicated by LOUT = 1 or 2 in line G1; the required data are entered in line G3) or if pumped outflow is selected (LOUT = 3 in line G1; the necessary data are entered in line G4). *

TSDEP(IS,MM) : Unit depth MM, feet.

TSAREA(IS,MM) : Surface area MM corresponding to TSDEP(IS,MM), square feet [square meters].

TSTORE(IS,MM) : Volume MM corresponding to TSDEP (IS,MM), cubic feet [cubic meters].

* If all values for volume are 0.0, volume is estimated with the depth-surface area data. *

TSQOU(IS,MM) : Outflow MM corresponding to TSDEP(IS,MM), cubic feet/second [cubic meters/second].

* This may be blank if LOUT = 1, 2, or 3 on line G1. *

---

*G3 LINE : DEPTH-OUTFLOW POWER EQUATION*

* OMIT THIS LINE IF LOUT = 0 OR 4 IN LINE G1. *

* The following three parameters represent the coefficients for the power equation depicting the depth-discharge relationship for the first outlet (there will be only one if LOUT = 1 in line G1). *

A1 : Depth-outflow equation coefficient. The units depend on exponent A2 (see below) and whether U.S. customary or metric units are used.

DO(1) : Depth-outflow equation minimum flow depth, feet [meters].

A2(1) : Depth-outflow equation exponent.

```
************************************************************************
*   The following four parameters represent the coefficients for the power   *
*   equation depicting the depth-discharge relationship for the second        *
*   outlet. These are only required if LOUT = 2 in line G1.                   *
************************************************************************
```

A1(2) : Depth -outflow equation coefficient. The units depend on exponent A2 (see below) and whether U.S. customary or metric units are used.

DO(2) : Depth-outflow equation minimum flow depth, feet [meters].

A2(2) : Depth-outflow equation exponent.

GEOM3 : Element to which outflow from second outlet is sent (specify the element number, i.e., NOE in the particular E1 line describing that element).

---

*G4 LINE : OUTFLOW PUMPING*

```
************************************************************************
*   OMIT THIS LINE IF LOUT = 0, 1, 2, OR 4 IN LINE G1.                        *
*                                                                             *
*   Two pumping rates, initiated at two different depths, are allowed for storage *
*   unit IS.                                                                  *
************************************************************************
```

TDSTAR(IS,1) : Depth at which TQPUMP(1) begins, feet [meters].

TDSTAR(IS,2) : Depth at which TQPUMP(2) begins, feet [meters]. This depth must be greater than or equal to TDSTAR(IS,1).

TQPUMP(IS,1) : Pumping rate when depth ≥ TDSTAR(IS,1), cubic feet/second [cubic meters/second].

TQPUMP(IS,2) : Pumping rate when depth ≥ TDSTAR(IS,2), cubic feet/second [cubic meters/second].

```
****************************************************
* If only one rate is desired, set this rate equal to the first   *
* rate.                                                            *
****************************************************
```

TDSTOP(IS) : Depth below which all pumping stops, feet [meters]. This must be less than or equal to the first depth, i.e., TDSTOP(IS) ≤ TDSTAR(IS).

---

*G5 LINE : INITIAL CONDITIONS IN INTERNAL STORAGE ELEMENT*

```
************************************************************************
*   This line enters the volume of water in storage unit IS at the start of the *
*   simulation, as well as the initial concentrations for the simulated quality *
*   constituents, if any.                                                     *
************************************************************************
```

STORL(IS) : Total volume of water in the storage unit at the start of the simulation, cubic feet [cubic meters].

```
*****************************************************************
*  The remaining parameters are only required if NPOLL > 0 in line B1 and STORL  *
*  > 0.0 (see above). The units must agree with those specified in the F1 lines.  *
*****************************************************************
```

PTCO(IS,1) : Concentration of constituent 1 in the storage unit at the start of the simulation.

PTCO(IS,2) : Concentration of constituent 2 in the storage unit at the start of the simulation. Required only if NPOLL ≥ 2 in line B1.

PTCO(IS,3) : Concentration of constituent 3 in the storage unit at the start of the simulation. Required only if NPOLL ≥ 3 in line B1.

PTCO(IS,4) : Concentration of constituent 4 in the storage unit at the start of the simulation. Required only if NPOLL = 4 in line B1.

---

*I1 LINE : NON-CONDUITS TRANSFERRED TO SUBSEQUENT BLOCK*

---

```
*****************************************************************
*  OMIT THIS LINE IF NOUTS = 0 IN LINE B1.                        *
*                                                                 *
*  Line I1 lists the non-conduit elements from which flow and water quality  *
*  constituents are transferred to subsequent blocks. A total of NOUTS (line  *
*  B1) elements are transferred. The maximum number of locations from which  *
*  this information may be taken is 100. The elements are identified by the  *
*  numbers assigned to them in the E1 lines (i.e., NOE).          *
*****************************************************************
```

JN(1) : First transfer location (element) number.

JN(2) : Second transfer location (element) number.

. .

. .

. .

JN(NOUTS) : Last transfer location (element) number.

---

*I2 LINE : CONDUITS WHERE DEPTHS ARE TO BE PRINTED*

---

```
*****************************************************************
*  Line I2 lists the conduit elements for which depths are printed. The number  *
*  of locations is indicated on this line (instead of line B1). The elements are  *
*  identified by the numbers assigned to them in the E1 lines (i.e., NOE).  *
*****************************************************************
```

NSURF : Number of elements for which depths will be printed.

JSURF(1) : First depth print location (element) number.

JSURF(2) : Second depth print location (element) number.

. .

. .

. .

JSURF(NSURF): Last depth print location (element) number.

*J1 LINE : NON-CONDUITS WHERE INPUT FLOWS AND WATER QUALITY CONSTITUENTS ARE TO BE PRINTED*

```
**************************************************************************
*   OMIT THIS LINE IF NNYN = 0 IN LINE B1.                               *
*                                                                        *
*   Line J1 lists the non-conduit elements for which input flows and water quality*
*   constituents are printed. The number of locations is indicated in line B1    *
*   (NNYN). The maximum number of locations is 80. The elements are identified by *
*   the numbers assigned to them in the E1 lines ( i.e., NOE). "Input" usually refers*
*   to data transferred from the RUNOFF Block.                           *
**************************************************************************
```

NYN(1) : First input location (element) number.

NYN(2) : Second input location (element) number.

. .

. .

NYN(NNYN) : Last input location (element) number.

*J2 LINE : NON-CONDUITS WHERE ROUTED FLOWS AND WATER QUALITY CONSTITUENTS ARE TO BE PRINTED*

```
**************************************************************************
*   OMIT THIS LINE IF NNPE = 0 IN LINE B1.                               *
*                                                                        *
*   Line J1 lists the non-conduit elements for which routed flows and water quality *
*   constituents are printed. The number of locations is indicated in line B1    *
*   (NNPE). The maximum number of locations is 80. The elements are identified *
*   by the numbers assigned to them in the E1 lines (i.e., NOE). The routed flows *
*   flows and constituents include those from upstream elements and those *
*   entered at that element.                                             *
**************************************************************************
```

NPE(1) : First routed flow location (element) number.

NPE(2) : Second routed flow location (element) number.

. .

. .

. .

NPE(NNPE) : Last routed flow location (element) number.

```
#########################################################################
#   IF INFILTRATION IS NOT SIMULATED, I.E., NINFIL = 0 IN LINE          #
#   B3, OMIT LINES K1 AND K2 AND SKIP TO LINE L1.                       #
#                                                                       #
#   The K1 and K2 lines are used to enter infiltration flows and constituent #
#   concentrations. The SWMM User's Manual provides some detail on how to #
#   estimate constant values for dry-weather infiltration, wet-weather infiltration, #
#   residual melting ice and frost infiltration, and ground water infiltration. The #
#   user is responsible for entering these values. The model only apportions #
#   the total infiltration flow over the entire sewer system based on conduit #
#   lengths and cross-sectional areas. The algorithm is definitely intended #
#   for single-event simulation.                                        #
#                                                                       #
#   As an alternative, the user may simply want to enter estimates of infiltration #
#   flow and constituent concentrations at several manholes with the E1 lines. #
#########################################################################
```

---

*K1 LINE : ESTIMATED INFILTRATION FLOWS*

---

```
*************************************************************************
*   The following values are infiltration flow totals for the entire sewer system. *
*   They will be apportioned over the system as described above.       *
*************************************************************************
```

DINFIL : Dry-weather infiltration, cubic feet/second [cubic meters/second].

GINFIL : Ground water infiltration, cubic feet/second [cubic meters/second].

RINFIL : Wet-weather infiltration, cubic feet/second [cubic meters/second].

RSMAX : Residual moisture peak, cubic feet/second [cubic meters/second].

```
*************************************************************************
*   The following constituent concentrations are applied uniformly throughout the *
*   system. NPOLL (line B1) concentrations are entered.                *
*************************************************************************
```

CPINF(1) : Concentration of quality constituent 1 (units according to NDIM, line F1).

. .

. .

. .

CPINF(NPOLL) : Concentration of quality constituent NPOLL (units according to NDIM, line F1).

---

*K2 LINE : MONTHLY DEGREE-DAYS*

---

```
*************************************************************************
*   Monthly degree-day data are used to determine when infiltration due to *
*   melting snow and ice (i.e., residual) is occurring. These are not needed if *
*   RSMAX = 0 in line K1, but zeros must still be entered.             *
*************************************************************************
```

NDD(1) : January degree-days, °F-day [no metric units].

. .

. .

. .

NDD(12) : December degree-days, °F-day [no metric units].

---

```
###########################################################################
#   Lines L1, L2, L3, M1, M2, M3, M4, N1, O1, O2, P1, and Q1 provide data for   #
#   a routine that estimates sewage flow from the residential, commercial, and  #
#   industrial areas in a watershed.                                            #
#                                                                               #
#   The amount of data and effort required is probably well beyond the benefit #
#   gained. Thus, this portion of the TRANSPORT Block is ignored in this guide. #
#   Sewage flows can be simulated by entering constant flows and constituent    #
#   concentrations at several manholes with the E1 lines. While this method does #
#   not allow for flow variations it is probably adequate for the large majority of #
#   applications.                                                               #
###########################################################################
```

---

*L1 LINE : FACTORS TO CORRECT YEARLY AVERAGE SEWAGE FLOWS TO AVERAGE DAILY SEWAGE FLOWS*

---

*L2 LINE : FACTORS TO CORRECT YEARLY AVERAGE BOD CONCEN TRATIONS TO AVERAGE DAILY BOD CONCENTRATIONS*

---

*L3 LINE : FACTORS TO CORRECT YEARLY AVERAGE SS* CONCEN*TRA-TIONS TO AVERAGE DAILY SS CONCENTRATIONS*

---

*M1 LINE : FACTORS TO CORRECT DAILY AVERAGE SEWAGE FLOWS TO HOURLY AVERAGE SEWAGE FLOWS*

---

*M2 LINE : FACTORS TO CORRECT DAILY AVERAGE BOD CONCENTRA-TIONS TO HOURLY AVERAGE BOD CONCENTRATIONS*

---

*M3 LINE : FACTORS TO CORRECT DAILY AVERAGE SS CONCENTRA-TIONS TO HOURLY AVERAGE SS CONCENTRATIONS*

---

*M4 LINE : FACTORS TO CORRECT DAILY AVERAGE COLIFORM CON-CENTRATIONS TO HOURLY AVERAGE COLIFORM CONCEN-TRATIONS*

---

*N1 LINE : GENERAL STUDY AREA DATA*

---

| | | |
|---|---|---|
| *O1 LINE* | : | *AVERAGE STUDY AREA DATA* |

| | | |
|---|---|---|
| *O2 LINE* | : | *CATEGORIZED STUDY AREA DATA* |

| | | |
|---|---|---|
| *P1 LINE* | : | *PROCESS FLOW CHARACTERISTICS* |

| | | |
|---|---|---|
| *Q1 LINE* | : | *SUBAREA DATA* |

| | | |
|---|---|---|
| *R1 LINES* | : | *HYDROGRAPH AND POLLUTOGRAPH INPUT* |

*************************************************************************
* The R1 lines are for direct input of flow and water quality concentrations at the non-conduits. They are omitted here since NCNTRL = 0 on line B3 and NINPUT = 0 on line B1. *
*************************************************************************

## A.4 EXTRAN BLOCK INPUT DATA

See NOTE at beginning of Section A.2.

| | | |
|---|---|---|
| *A1 LINE* | : | *TITLE LINES* |
| TITLE | : | Two lines with heading to be printed on output. |

*****************************************************
* There must be two lines, each in single quotes ('Run No. 1'). *
*****************************************************

##########################################################################
\# The latest versions of EXTRAN contain two lines (B0 and BB) with some options which do little to enhance model performance (in fact, some have not always worked very well — although they may have been improved in the very recent past). They may be omitted and so they are here. #
##########################################################################

| | | |
|---|---|---|
| *B0 LINE* | : | *MISCELLANEOUS CONTROL PARAMETERS I* |

| | | |
|---|---|---|
| *BB LINE* | : | *MISCELLANEOUS CONTROL PARAMETERS II* |

| | | |
|---|---|---|
| *B1 LINE* | : | *CONTROL DATA* |

*************************************************************************
* This line sets some of the run control parameters and EXTRAN Block options, as do lines B2 and B3. *
*************************************************************************

NTCYC : Number of time steps desired.

```
****************************************************
* There is no limit on the number of time steps, and thus the *
* length of the simulation. However, the EXTRAN Block is not *
* well designed, in terms of its output, for long-term continuous*
* simulation. The user needs to use the print options in this *
* line and in lines B3 through B8 carefully. *
****************************************************
```

DELT : Length of time step, seconds.

```
****************************************************
* The stability of the EXTRAN Block is sensitive to the size *
* of the time step. The guidance in the SWMM User's *
* Manual Addendum is fairly good. However, a good rule of *
* thumb is 5 to 15 seconds. *
* *
* Stability problems manifest themselves in unusually erratic *
* flows and/or large continuity errors (an analysis of continuity *
* is presented in the EXTRAN output). *
****************************************************
```

TZERO : Starting time of the simulation, decimal hours (e.g., 5:30 p.m. is 17.5).

NSTART : First time-step to begin print cycle.

```
****************************************************
* This will delay the printing of detailed "intermediate" output *
* until time step NSTART. This is output that reports the *
* status of the drainage system at all junctions and conduits. *
* The output can be voluminous (and not very easy to decipher). *
* Use NSTART and INTER below to judiciously create this *
* kind of output. "Intermediate" output is most useful in early *
* runs when the viability of the model is still in question. After *
* the model is running well the "intermediate" output *
* becomes less important. *
****************************************************
```

INTER : Number of time steps between intermediate print cycles during the simulation. The number of cycles printed is (NTCYC - NSTART)/ INTER.

```
****************************************************
* See comment for NSTART. *
****************************************************
```

JNTER : Number of time steps between time-history summary print cycles at the end of the simulation. The number of cycles printed is NTCYC/JNTER.

```
****************************************************
* Time-history summaries of head and flow are printed at the *
* end of the simulation for the junctions and conduits specified *
* in lines B3 through B8. An appropriate value would be one *
* that insures that the time history is accurately depicted *
* without having to print every time step. A good rule of thumb *
* is to set JNTER at 10. The time step size is often less than *
* 30 seconds. At 30 seconds a value of 10 would have the program *
* print flows and heads every 5 minutes. *
****************************************************
```

REDO : Hot-start file manipulation parameter.
= 0, no hot-start file is created or used.
= 1, read NSCRAT(2) for initial flows, heads, areas, and velocities.
= 2, create a new hot-start file on NSCRAT(2).
= 3, create a new hot-start file but use the fold file as the initial conditions. The old file is subsequently erased and a new file created.

```
****************************************************
* The "hot-start" option can be used as a way to let the *
* program establish initial flow and head conditions in the *
* drainage system. *
* *
* Initial conditions in combined sewers can be developed by *
* beginning a model run with the system initially "dry" and *
* allowing dry-weather (sanitary) flows to reach equilibrium in *
* the system. Dry-weather flows can be entered directly at *
* various junctions. After this equilibrium is reached, the *
* wet-weather flow portion of the simulation can begin. This *
* can be accomplished in one run as long as enough dry- *
* weather time elapses before introducing the runoff flows. *
* This "warm-up" period is seldom long. *
* *
* This option allows the user to save the results of a "warm- *
* up" run, so as to avoid re-executing that portion in subsequent *
* runs. It is not very cumbersome to simply add a little time at *
* the beginning of each run for the "warm-up", but the user may *
* want to take advantage of the "hot-start" option anyway. Either *
* way, however, is easier than entering the initial conditions in *
* lines C1 and D1 (parameters QO and YO, respectively). *
****************************************************
```

*B2 LINE : MORE CONTROL DATA*

METRIC : Metric input-output.
= 0, use U.S. customary units.
= 1, use metric units. Metric input is indicated in brackets [ ] through the remainder of this guide.

NEQUAL : Modify short pipe lengths using an equivalent pipe to ease time step limitations. Refer to the SWMM User's Manual Addendum.
= 0, do not modify.
= 1, modify short pipe lengths.

```
****************************************************
* Essentially, the time step size needs to be small enough to   *
* ensure that the propagation of flow changes through a conduit*
* are properly simulated. A long time step may not be able to *
* keep up with rapid changes in flow and head (see Chapter 2). *
*                                                                *
* One way to overcome this problem is to replace short pipes *
* with hydraulically equivalent, longer pipes. The user can do *
* this directly or rely on the program to carry out the necessary *
* calculations (NEQUAL = 1). Either way will work, but the    *
* automatic method is more convenient.                          *
****************************************************
```

AMEN : Surface area for all manholes, square feet [square meters]. Used for surcharge calculations. (The default value is based on a manhole diameter of 4 feet [1.22 meters]).

ITMAX : Maximum number of iterations to be used in the numerical method used for surcharge calculations (30 is recommended for ISOL = 0 or ISOL = 1; 10 is recommended for ISOL = 2).

```
****************************************************
* Line B0 was ignored earlier. One of its parameters, ISOL,   *
* allows for different numerical solution techniques other than the *
* explicit method (ISOL = 0) described in the SWMM User's *
* Manual Addendum. Experience has shown that the explicit  *
* method is usually quite adequate. Set ITMAX = 30. If        *
* convergence problems occur, increasing ITMAX may help.  *
*                                                                *
* Convergence problems can be easily uncovered by inspecting *
* the "intermediate" detailed output. Section 4 in the SWMM    *
* User's Manual Addendum effectively presents methods for   *
* dealing with convergence problems.                            *
****************************************************
```

SURTOL : Fraction of the average flow in surcharged areas to be used as convergence criterion for surcharge iterations (0.05 is recommended for ISOL = 0 or 1; 0.0025 for ISOL = 2).

```
****************************************************
* See comments for ITMAX. Set SURTOL = 0.05. Refer to    *
* Section 4 in the SWMM User's Manual Addendum if         *
* convergence problems occur.                                  *
****************************************************
```

| *B3 LINE* | : | *MORE CONTROL DATA* |
|---|---|---|
| NHPRT | : | Number of junctions for printing of time-history of head. There is a limit on the number of junctions for which this can occur (which varies from one SWMM release to another). These junctions are identified in line B4. |
| NQPRT | : | Number of conduits for printing of time-history of flow. There is a limit on the number of conduits for which this can occur (which varies from one SWMM release to another). These conduits are identified in line B5. |
| NPLT | : | Number of junction for plotting of time-history of head. There is a limit on the number of junctions for which this can occur (which varies from one SWMM release to another). These junctions are identified in line B6. |
| LPLT | : | Number of conduits for plotting of time-history of flow. There is a limit on the number of conduits for which this can occur (which varies from one SWMM release to another). These conduits are identified in line B7. |
| NJSW | : | Number of input junctions at which user-supplied input hydrographs are used. |

**************************************************
* It will be assumed here that the user will have the RUNOFF*
* Block supply input to the EXTRAN Block. Set NJSW = 0. *
**************************************************

########################################################################
\# Lines B4, B5, B6, B7, and B8 are used to specify the locations for which #
\# detailed time histories of head and flow are to be printed. #
########################################################################

| *B4 LINE* | : | *PRINTED JUNCTION HEADS* |
|---|---|---|

*****************************************************************
* OMIT THIS LINE IF NHPRT = 0 IN LINE B3. *
* *
* This line is used to indicate the NHPRT junctions for which detailed time histories *
* of head are to be printed. The junctions must be listed in the D1 lines (and *
* exact matches found for the identifiers). If names are used to identify the *
* junctions they must be placed in single quotes (e.g., 'JUNC 1'). *
*****************************************************************

| JPRT(1) | : | First junction number/name for detailed printing of junction heads. |
|---|---|---|
| JPRT(2) | : | Second junction number/name for detailed printing of junction heads. |

.          .

.          .

.          .

JPRT(NHPRT) : Last junction number/name for detailed printing of junction heads.

---

*B5 LINE : PRINTED CONDUIT FLOWS*

```
*****************************************************************
*  OMIT THIS LINE OF NQPRT = 0 IN LINE B3.                      *
*                                                               *
*  This line is used to indicate the NQPRT conduits for which detailed time histories*
*  of flow are to be printed. The conduits must be listed in the C1 lines (and   *
*  exact matches found for the identifiers). If names are used to identify the   *
*  conduits they must be placed in single quotes (e.g., 'PIPE 1').              *
*****************************************************************
```

CPRT(1) : First conduit number/name for detailed printing of conduit flows.

CPRT(2) : Second conduit number/name for detailed printing of conduit flows.

.          .

.          .

.          .

CPRT(NQPRT) : Last conduit number/name for detailed printing of conduit flows.

---

*B6 LINE : PLOTTED JUNCTION HEADS*

```
*****************************************************************
*  OMIT THIS LINE IF NPLT = 0 IN LINE B3.                       *
*                                                               *
*  This line is used to indicate the NPLT junctions for which detailed time     *
*  histories of head are to be plotted. The junctions must be listed in the D1  *
*  lines (and exact matches found for the identifiers). If names are used to    *
*  identify the junctions they must be placed in single quotes (e.g., 'JUNC 1'). *
*****************************************************************
```

JPLT(1) : First junction number/name for detailed plotting of junction heads.

JPLT(2) : Second junction number/name for detailed plotting of junction heads.

.          .

.          .

.          .

JPLT(NPLT) : Last junction number/name for detailed plotting of junction heads.

*B7 LINE : PLOTTED CONDUIT FLOWS*

***********************************************************************
* OMIT THIS LINE IF LPLT = 0 IN LINE B3.JPRT(1) *
* *
* This line is used to indicate the LPLT conduits for which detailed time histories *
* of flow are to be plotted. The conduits must be listed in the C1 lines (and *
* exact matches found for the identifiers). If names are used to identify the *
* junctions they must be placed in single quotes (e.g., 'PIPE 1'). *
***********************************************************************

CPRT(1) : First conduit number/name for detailed plotting of conduit flows.

CPRT(2) : Second conduit number/name for detailed plotting of conduit flows.

. .

. .

. .

CPRT(NQPRT) : Last conduit number/name for detailed plotting of conduit flows.

*B8 LINE : PLOTTED CONDUIT UPSTREAM/DOWNSTREAM ELEVATIONS*

***********************************************************************
* Line B8 lists the <u>conduits</u> for which elevations are to be printed. The number *
* of locations is indicated on this line, instead of line B3. The upstream and *
* downstream elevations are placed on the same plot. <u>This line is omitted if</u> *
* <u>these plots are not desired</u>. The conduits must be listed in the C1 lines (and *
* exact matches found for the identifiers). If names are used to identify the *
* conduits they must be placed in single quotes (e.g., 'PIPE 1'). *
***********************************************************************

NSURF : Number of conduits for which elevations will be plotted.

JSURF(1) : First conduit number/name for plot of upstream/downstream elevations.

JSURF(2) : Second conduit number/name for plot of upstream/downstream elevations.

. .

. .

. .

JSURF(NSURF): Last conduit number/name for plot of upstream/downstream elevations.

*C1 LINES : CONDUIT DATA*

***********************************************************************
* This line is repeated for each conduit in the modeled sewer system. There *
* is a limit on the number of conduits (which varies from one SWMM *
* release to another), but it may be increased (see the SWMM User's Manual). *
***********************************************************************

NCOND : Conduit number (any valid integer, but more than 5 digits is discouraged) or conduit name (enclose in single quotes).

NJUNC(1) : Junction number at upstream end of conduit or junction name (enclose in single quotes). Junction must be listed in the D1 lines.

NJUNC(2) : Junction number at downstream end of conduit or junction name (enclose in single quotes). Junction must be listed in the D1 lines.

QO : Initial flow, cubic feet/second [cubic meters/second].

```
**************************************************
* See comments for REDO in line B1.              *
**************************************************
```

```
*****************************************************************
*  The remaining parameters are used to describe the physical characteristics of  *
*  each conduit. The type is specified with NKLASS. The rest provide more          *
*  detailed information. Note that the same parameter may have different           *
*  meanings depending on the type of conduit.                                       *
*****************************************************************
```

NKLASS : Type of conduit shape.
= 1, circular.
= 2, rectangular.
= 3, horseshoe.
= 4, eggshape.
= 5, baskethandle.
= 6, trapezoidal channel.
= 7, parabolic/power function channel.
= 8, irregular (natural) channel.

```
**************************************************
* The regularly shaped conduits are adequate for vast majority*
* of applications.  Throughout the remainder of this guide,   *
* NKLASS = 8 will be ignored. Do not set NKLASS = 8.          *
**************************************************
```

AFULL : Cross-sectional area of conduit, square feet [square meters]. Required only for NKLASS = 3, 4, or 5. (Geometric properties for types 3 to 5 may be found in Section 6 of the SWMM User's Manual.)

DEEP : Vertical depth of conduit, feet [meters]. (This is the diameter of a circular pipe.)

OR

Maximum depth of a parabolic/power function channel, feet [meters].

WIDE : Maximum width of conduit types (NKLASS = 2, 3, 4, or 5), feet [meters]. Not required for circular pipes (NKLASS = 1).

OR

Bottom width for trapezoidal channel (NKLASS = 6), feet [meters}.

OR

Top width for parabolic/power function channel (NKLASS = 7), feet [meters].

LEN : Length of conduit, feet [meters].

*********************************************************************
* The following two parameters place the invert of the conduit relative to the *
* inverts of the upstream and downstream junctions. <u>At least one</u> of the *
* conduits joining a junction must have an invert identical to the invert of the *
* junction (i.e., ZP must be 0.0 for one of the conduits). If none do this in fact, *
* set the junction invert to the invert of the lowest conduit joining the junction *
* (see D1 lines). *
*********************************************************************

ZP(1) : Distance of conduit invert above junction invert (at NJUNC(1), feet [meters].

ZP(2) : Distance of conduit invert above junction invert (at NJUNC(2), feet [meters].

ROUGH : Manning's coefficient (includes entrance, exit, expansion, and contraction losses).

STHETA : Slope of one side of trapezoidal channel (horizontal/vertical; set STHETA = 0.0 for vertical walls). Required only for NKLASS = 6.

SPHI : Slope of other side of trapezoid channel (horizontal/vertical; set STHETA = 0.0 for vertical walls). Required only for NKLASS = 6.

---

#######################################################################
# The C2, C3, and C4 lines are used to enter hydraulic data for natural channels#
# (NKLASS = 8 in line C1). This option is rarely used and, thus, these lines #
# are omitted here. #
#######################################################################

*C2 LINE : NATURAL CHANNEL ROUGHNESS DATA*

*C3 LINE : NATURAL CHANNEL CROSS SECTION DATA*

*C4 LINE : NATURAL CHANNEL PROFILE DATA*

*D1 LINES : JUNCTION DATA*

******************************************************************
* This line is repeated for each junction in the modeled sewer system. There is *
* a limit on the number of junctions (which varies from one SWMM release to *
* another), but it may be increased (see the SWMM User's Manual). *
******************************************************************

JUN : Junction number (any valid integer, but more than 5 digits is discouraged) or junction name (enclose in single quotes).

GRELEV : Ground elevation, feet [meters].

Z : Invert elevation, feet [meters].

****************************************************
* See comments for ZP(1) and ZP(2) in line C1. *
****************************************************

QINST : Net constant flow into junction, cubic feet/second [cubic meters/second]. A positive value indicates an inflow. A negative value indicates a withdrawal or loss.

****************************************************
* This is a very useful parameter. It allows the direct input of *
* flows to the junction. While it is limited to a constant flow, *
* QINST is very commonly used to simulate sanitary sewage *
* flows to the system. For most applications this is *
* adequate. *
****************************************************

YO : Initial depth above junction invert elevation, feet [meters].

****************************************************
* See comments for REDO in line B1. *
****************************************************

*E1 LINES : STORAGE JUNCTION DATA*

******************************************************************
* This line is repeated for each storage junction in the modeled sewer system. *
* There is a limit on the number of storage junctions (which varies from one SWMM *
* release to another), but it may be increased (see the SWMM User's Manual). *
* *
* IMPORTANT: Each storage junction must also have been entered with the *
* junction data (D1 lines). *
* *
* There are three types of storage junctions: *
* (1) Constant surface area with depth (stage). Only JSTORE, ZTOP, and *
* ASTORE are required in line E1. Line E2 is omitted. *

```
*   (2) Variable surface area with depth described by a set of data pairs. JSTORE, *
*       ZTOP, ASTORE, and NUMST are required in line E1. NUMST values            *
*       in line E2 are necessary.                                                 *
*   (3) Variable surface area with depth described by a power function. JSTORE,*
*       ZTOP, ASTORE, and NUMST are required in line E1. There should          *
*       be two data values in line E2.                                            *
*******************************************************************************
```

JSTORE : Junction number containing storage unit or junction name (enter in single quotes).

ZTOP : Junction crown elevation (must be higher than crown of highest pipe connected to the storage junction), feet [meters].

ASTORE : Storage volume per foot (or meter) of depth for a constant surface area storage junction, cubic feet/foot [cubic meters/meter].

OR

Set ASTORE(J) < 0 to indicate a variable surface area storage junction or if power function will be used to describe the storage-depth relationship.

NUMST : Total number of stage/storage area points in the E2 lines if ASTORE < 0 (NUMST must be < 30).

OR

Enter a value of –2 if a power equation is used to describe the surface area-depth relationship.

Not required if constant area junction is desired.

---

*E2 LINES : VARIABLE -AREA STORAGE JUNCTION DATA*

```
*******************************************************************************
*   THIS LINE FOLLOWS LINE E1 EACH TIME ASTORE < 0 IN LINE E1,   *
*   OTHERWISE OMIT.                                                           *
*                                                                             *
*   This line is repeated for each storage junction characterized by a variable    *
*   surface area-depth relationship, whether through data pairs or a power          *
*   equation.                                                                 *
*******************************************************************************
```

QCURVE(1,1) : Surface area of storage junction at the first depth, QCURVE (2,1), acres [hectares].

OR

If NUMST in line E1 equals –2 this is the coefficient of the power function describing the surface area-depth relationship.

```
*****************************************************
* The power equation option assumes that the surface area is *
* in square feet [square meters] as opposed to the data pair *
* option which assumes the surface area is in acres [hectares].*
*****************************************************
```

QCURVE(2,1) : First depth above junction invert, feet [meters].

OR

If NUMST equals –2 this is the exponent of the power function and is the last value entered.

. .

. .

. .

QCURVE(1,NUMST): Surface area of storage junction at the last depth, QCURVE (2,NUMST), acres [hectares].

QCURVE(2,NUMST): Last depth above junction invert, feet [meters].

---

*F1 LINES : ORIFICE DATA*

```
*****************************************************************
* This line is repeated for each junction containing an orifice in the modeled *
* sewer system. There is a limit on the number of orifice junctions (that varies *
* from release to release), but it may be increased (see the SWMM User's Manual). *
*                                                                               *
* IMPORTANT: Each orifice junction must also have been entered with             *
* the junction data (D1 lines).                                                 *
*****************************************************************
```

NJUNC(1) : Junction number containing an orifice or junction name (enter in single quotes).

NJUNC(2) : Junction number to which orifice discharges or junction name (enter in single quotes).

NKLASS : Type of orifice.
= 1, side outlet.
= 2, bottom outlet.
= –1, time-history side outlet orifice with data entered in line F2.
= –2, time-history bottom outlet orifice, with data entered on data group F2.

```
*****************************************************
* If NKLASS = –1 or –2, an orifice with time-varying char- *
* acteristics is modeled. This is an not likely to be used. *
* It is omitted here. Set NKLASS = 1 or 2.                  *
*****************************************************
```

AORIF : Orifice area, square feet [square meters].

CORIF : Orifice discharge coefficient.

ZP : Distance of orifice invert above junction floor, feet [meters]. Required only for side outlet orifices (NKLASS = 1).

---

*F2 LINES : TIME-HISTORY ORIFICE DATA*

---

****************************************************************
* The time-history orifice was not selected in line F1, i.e., NKLASS was *
* set to either 1 or 2. Therefore, this line is omitted here. *
****************************************************************

---

*G1 LINES : WEIR JUNCTION DATA*

---

****************************************************************
* This line is repeated for each junction containing a weir in the modeled *
* sewer system. There is a limit on the number of weir junctions (which *
* varies from one SWMM release to another), but it may be increased (see the *
* SWMM User's Manual). *
* *
* IMPORTANT: Each weir junction must also have been entered with the *
* junction data (D1 lines). *
* *
* Weirs are a constant source of problems in EXTRAN, especially when sur- *
* charging occurs. It is sometimes helpful to model them with an equivalent *
* pipe. See the SWMM User's Manual Addendum. *
****************************************************************

NJUNC(1) : Junction number at which weir is located or junction name (enter in single quotes).

NJUNC(2) : Junction number to which weir discharges or junction name (enter in single quotes).

OR

To designate this weir as an outfall weir, set NJUNC(N,2) = 0 or ' ' (one space between quotes).

KWEIR : Type of weir.
= 1, transverse weir.
= 2, transverse weir with tide gate.
= 3, side flow weir.
= 4, side flow weir with tide gate.

YCREST : Height of weir crest above invert, feet [meter].

YTOP : Height to top of weir opening above invert (surcharge level), feet [meters].

WLEN : Weir length, feet [meters].

COEF : Coefficient of discharge for weir.

---

*H1 LINES : PUMP JUNCTION DATA*

---

```
*******************************************************************
*  This line is repeated for each junction containing a pump in the modeled  *
*  sewer system. There is a limit on the number of pump junctions (which     *
*  varies from one SWMM release to another), but it may be increased         *
*  (see the SWMM User's Manual).                                              *
*                                                                             *
*  IMPORTANT: Each pump junction must also have been entered with the        *
*  junction data (D1 lines). In addition, only one conduit may be connected  *
*  to a pump junction.                                                        *
*******************************************************************
```

IPTYP : Type of pump.
= 1, off-line pump with wet well (program will set pump junction invert to –100).
= 2, in-line lift pump.
= 3, three-point head-discharge pump curve.

NJUNC(1) : Junction number being pumped or junction name (enter in single quotes).

NJUNC(2) : Pump discharge goes to this junction (indicate with a number or name entered in single quotes).

PRATE(1) : Lower pumping rate, cubic feet/second [cubic meters/second].

PRATE(2) : Mid-pumping rate, cubic feet/second [cubic meters/second].

PRATE(3) : High pumping rate, cubic feet/second [cubic meters/second].

VRATE(1) : If IPTYP = 1, enter the wet well volume for mid-rate pumps to start, cubic feet [cubic meters].

OR

If IPTYP = 2, enter the junction depth for mid-rate pumps to start, feet [meters].

OR

If IPTYP = 3, enter the head difference (head at junction down stream of pump minus head at junction upstream of pump) associated with the lowest pumping rate, feet [meters]. (This will be the highest head difference.)

VRATE(2) : If IPTYP = 1, enter the wet well volume for high-rate pumps to start, cubic feet [cubic meters].

OR

If IPTYP = 2, enter the junction depth for high-rate pumps to start, feet [meters].

OR

If IPTYP = 3 enter the head difference associated with the mid-pumping rate, feet [meters].

```
*****************************************************************
*   The following two parameters are required only if IPTYP = 1 or 3.   *
*****************************************************************
```

VRATE(3) : If IPTYP = 1, enter total wet well capacity, cubic feet [cubic meters].

OR

If IPTYP = 3, enter the head difference associated with the highest pumping rate, feet [meters]. (This will be the lowest head difference.)

VWELL : If IPTYP = 1, then enter the initial wet well volume, cubic feet [cubic meters].

OR

If IPTYP = 3, then enter the initial depth in the pump inflow junction, feet [meters].

```
*****************************************************************
*   The following two parameters are required only if IPTYP = 2 or 3.   *
*****************************************************************
```

PON : Depth of water in the pump inflow junction to turn pump on, feet [meters].

POFF : Depth of water in the pump inflow junction to turn pump off, feet [meters].

---

*I1 LINES : OUTFALL (WITHOUT TIDEGATE) JUNCTION DATA*

```
*****************************************************************
*   This line is repeated for each junction containing an outfall, without a tidegate,  *
*   in the modeled sewer system. There is a limit on the number of outfall   *
*   junctions (which varies from one SWMM release to another), but it may be   *
*   increased (see the SWMM User's Manual).   *
*   *
*   IMPORTANT: Each out fall junction must also have been entered with the   *
*   junction data (D1 lines). In addition, only one conduit may be connected to   *
*   an outfall junction.   *
*****************************************************************
```

JFREE : Number or name of outfall junctions with out tide gate (no back-flow restriction).

NBCF : Type of boundary condition from the sequence of J1, J2, J3, and J4 line groups below.

---

*I2 LINES : OUTFALL (WITH TIDEGATE) JUNCTION DATA*

```
*************************************************************************
*   This line is repeated for each junction containing an outfall, with a tidegate,   *
*   in the modeled sewer system. There is a limit on the number of outfall            *
*   junctions (which varies from one SWMM release to another), but it may be          *
*   increased (see the SWMM User's Manual).                                           *
*                                                                                     *
*   IMPORTANT: Each outfall junction must also have been entered with the             *
*   junction data (D1 lines). In addition, only one conduit may be connected          *
*   to an outfall junction.                                                           *
*************************************************************************
```

JGATE : Number or name of outfall junction with tide gate (Back-flow restriction).

NBCG : Type of boundary condition from the sequence of J1, J2, J3, and J4 line groups below.

---

```
##########################################################################
#   Repeat the J1, J2, J3, and J4 group of lines for up to 20 different sets of       #
#   boundary conditions. J2, J3, or J4 may not be required, depending on the          #
#   option selected in J1. The order in which these groups appear determines          #
#   the value for NBCF and NBCG in lines I1 and I2. For example, if the third         #
#   group of J1-J4 cards is the set of boundary conditions desired for a particular   #
#   non-gate outfall, NBCF = 3.                                                       #
##########################################################################
```

*J1 LINE : BOUNDARY CONDITION INDEX FOR CONDITION SET I*

NTIDE : Boundary condition index.

= 1, no water surface at outfalls (elevated discharge). Lines J2, J3, and J4 lines are not required.

= 2, controlling water surface at outfall is at constant elevation A1 (entered in line J2), feet [meters].

= 3, tide coefficients (entered in line J2) provided by user.

= 4, program will compute tide coefficients (requires lines J2, J3, and J4).

= 5, stage-history of water surface elevations input by user (re quires lines J3 and J4). Program uses linear interpolation between data points.

---

*J2 LINE : STAGE AND/OR TIDAL COEFFICIENTS FOR CONDITION SET I*

**********************************************************************
* OMIT THIS LINE IF NTIDE = 1 OR 5 ON DATA GROUP J1. *
**********************************************************************

A1 : First tide coefficient, feet [meters].
OR
Constant elevation (if NTIDE = 2 on line J1), feet [meters].

W : Tidal period, hours. Required only if NTIDE(I) = 3 or 4 in line J1, otherwise omit.

**********************************************************************
* The next six fields are required only if NTIDE = 3 on line J1, otherwise *
* omit. Refer to the SWMM User's Manual Addendum for definitions of the *
* coefficients. *
**********************************************************************

A2 : Second tide coefficient, feet [meters].

A3 : Third tide coefficient, feet [meters].

A4 : Fourth tide coefficient, feet [meters].

A5 : Fifth tide coefficient, feet [meters].

A6 : Sixth tide coefficient, feet [meters].

A7 : Seventh tide coefficient, feet [meters].

*J3 LINE : TIDAL INFORMATION FOR CONDITION SET I*

**********************************************************************
* OMIT THIS LINE IF NTIDE = 1, 2, OR 3 ON DATA GROUP J1. *
**********************************************************************

KO : Type of tidal input. Not required if NTIDE = 5.
= 0, input is in the form of a time series of NI tidal heights.
= 1, input is in the form of the high and low water values found in the tide tables (HHW, LLW, LHW, and HLW). NI (immediately below) must be 4.

NI : Number of information points.

NCHTID : Tide information print control.
= 0 Do not print information.
= 1 Print information on tide coefficients or stage history.

DELTA : Convergence criterion (default = 0.0005) for fitting the tidal function, feet [meters]. Not required if NTIDE = 5.

*J4 LINES : TIME-TIDAL STAGE INFORMATION FOR CONDITION SET I*

```
*****************************************************************
*   OMIT THIS LINE IF NTIDE = 1, 2, OR 3 ON DATA GROUP J1.          *
*                                                                   *
*   Enter five pairs of time and stage information per line until NI (line J3)   *
*   points are listed. Include the line identifier, J 4, on each line.          *
*****************************************************************
```

TT(1) : Time of day, first information point, hours.

YY(1) : Tidal stage at time TT(1), feet [meters].

.
.
.

TT(NI) : Time of day, information point NI, hours. Increase hours past 24 if necessary.

YY(NI) : Tidal stage at time TT(NI), feet [meters].

```
#####################################################################
#   Lines K1, K2, and K3 are used for the direct input of hydrographs at various  #
#   locations in the drainage system. They are omitted because NJSW = 0 in line B3. #
#####################################################################
```

*K1 LINE : USER INPUT HYDROGRAPHS*

*K2 LINES : HYDROGRAPH NODES*

*K3 LINES : HYDROGRAPHS*

# INDEX